RIBA
HEALTH AND SAFETY GUIDE

Second edition

© RIBA Publishing, 2024

Published by RIBA Publishing, 66 Portland Place, London, W1B 1AD

ISBN 978 1 91572 201 0

British Library Cataloguing-in-Publication Data
A catalogue record for this book is available from the British Library.

Commissioning Editor: Alex White
Production: Richard Blackburn
Designed by Kneath Associates
Typeset by Fakenham Prepress Solutions
Printed and bound by TJ Books, Cornwall
All illustrations, including cover, by Natalie Gall

www.ribapublishing.com

MIX
Paper from
responsible sources
FSC
www.fsc.org
FSC® C013056

RIBA
HEALTH AND SAFETY GUIDE

Second edition

Dieter Bentley-Gockmann

RIBA Publishing

CONTENTS

ABOUT THE AUTHOR

Dieter Bentley-Gockmann is a Chartered Architect with experience of working for small and medium-sized practices involving the design and construction of a diverse portfolio of projects including residential, commercial and mixed-use developments. Dieter graduated from the Welsh School of Architecture, Cardiff University in 1995, completed a Master's degree in Construction Law and Arbitration at King's College, London in 2005 and since 2006 has been Director, Legal and Technical Services at EPR Architects Limited.

In his role as Director of EPR Architects, Dieter oversees the practice's organisational capability to deliver projects, including higher-risk buildings, in accordance with the relevant requirements of the Building Safety Act 2022 and the Building Regulations 2010. This includes advising the practice's designers about all aspects of regulatory and professional compliance and individual competence. In particular, he ensures that designers, including those acting in the capacity of principal designer, are aware of, understand, and have the appropriate skills, knowledge, experience and behaviours to discharge their duties under the Construction (Design and Management) Regulations 2015 and The Building Regulations etc. (Amendment) (England) Regulations 2023.

Dieter is a former chair of the RIBA's Regulations and Standards Expert Advisory Group and current member of the RIBA's Fire Safety Expert Advisory Group and the RIBA's Principal Designer Register advisory steering group.

Dieter is the author of the *RIBA Principal Designer's Guide*, a companion guide to the *Health and Safety Guide*, that provides guidance for architects and designers undertaking principal designer duties.

ACKNOWLEDGEMENTS AND DEDICATION

I would like to acknowledge the input and guidance of Adrian Dobson, Alex Tait and Alex Ryan at the RIBA, my many colleagues at EPR Architects and across the profession, in particular Sarah Susman and Paul Bussey, and Alex White, Caroline Ellerby and Richard Blackburn at RIBA Publishing, and thank them all for their invaluable insight, support and assistance in preparing the second edition of this guide. Thank you, as well, to Natalie Gall for her wonderful illustrations.

In particular, I would like to thank my husband, Neil, for his continued support, encouragement and belief in the value my contribution in writing this guide makes to the vital work being undertaken by so many others to improve our profession, the construction industry and the built environment.

Finally, I would like to dedicate this edition to Sam Webb MBE, whose critical role and lifetime contribution to the campaign for improvements in building safety is an inspiration and an exemplary role model for all of us as we strive to be better architects and designers.

Dieter Bentley-Gockmann
EPR Architects
August 2023

INTRODUCTION

As architects, we enjoy the privilege of being able to help shape the world around us. We have the opportunity to build communities and positively impact on the daily lives of all those that interact with our work through collaboration with our clients and their design and construction teams. As designers, lead designers and principal designers, we have an opportunity to influence the decision-making processes of our clients and their design and construction teams.

With these opportunities comes a responsibility to see that any decisions over which we exert control or influence are made bearing in mind the best interests of everyone that will be affected by our projects. In particular, we have a responsibility to ensure that every person that comes into contact with our projects, from inception through to occupation, maintenance, adaptation, conversion or subsequent demolition, is able to do so safely and with no detrimental effect on their health or wellbeing. This includes a duty to keep ourselves and our colleagues safe.

The responsibility to always be mindful of safety issues and create architecture that safeguards the wellbeing of the public goes to the core of everything that we do as architects; be that as a student, academic, researcher, practitioner, architectural assistant, project runner, practice principal, sole practitioner, employed or self-employed architect.

None of us set out to put ourselves at risk, design an unsafe project or knowingly cause harm. We must be conscious, however, that in our often highly pressured industry, with increasingly complex design, procurement and project delivery processes, we risk being distracted from paying sufficient attention to the elements of our professional services that ensure our projects are – and remain – safe.

The forthcoming RIBA health and safety test for architects is for everyone working and studying across the profession, and this companion guide has been produced by architects for architects, to be a study aid for architectural students and an aide-memoire for qualified architects preparing for the forthcoming test. Perhaps, more importantly, this guide is intended to remind us all that safe practice and design should be at the heart of what we do. We all have a duty to act professionally at all times and we all have a responsibility to ensure that everyone affected by the built environments we create is safe and feels safe.

Since the first edition of this guide was published, we have witnessed a growing momentum in the work being done by numerous professional institutes, organisations and individuals across the industry to improve building safety and the competence of all those involved in the design, procurement and construction of buildings, including higher-risk buildings. This change has been motivated in no small part by the enactment of the Building Safety Act 2022 and the secondary legislation introduced under it.

The second edition of this guide has been expanded and updated to include guidance for architects and designers on the Building Safety Act. In particular, it includes details on the duties and competence requirements introduced by the Act relevant to designers, and the procedural requirements relating to the building control regime applicable to buildings in England, including the regulated gateway approval regime for higher-risk buildings.

Chapter 1 considers how you should prepare for every site visit, whether to an unoccupied or occupied site, to ensure you are adequately equipped prior to commencing your visit. Chapter 2 details how you should conduct your site visit, whether to an unoccupied or occupied site, to ensure you are able to undertake your visit safely. The typical and more significant hazards you might face during a site visit are covered in Chapter 3, along with guidance on how you might consider managing these to ensure you remain safe. In Chapter 4 we explore the principles of design risk management, the general principles of prevention and the role of effective communication and co-ordination. Chapter 5 looks at the minimum standards with which you need to be familiar, including statutes and the regulatory environment, non-statutory guidance and codes of conduct. Chapter 6 addresses the legal duties imposed on designers and principal designers under the CDM Regulations. Chapter 7 details the statutory duties imposed on designers and the relevant changes to the building control regime introduced under the Building Safety Act 2022, including changes relevant to the design and approval of higher-risk buildings. Chapter 8 provides guidance on general aspects of building safety design and the minimum competence criteria that designers need to be familiar with relating to building design, including the design of higher-risk buildings. Finally, Chapter 9 covers the principles of fire science, the fire performance of construction materials and aspects of fire safety design.

CHAPTER 1:
PREPARING
TO VISIT SITE

Undertaking site visits is vital to our role as architects and fulfils several functions, from enabling us to understand and analyse the context within which our projects will reside, to inspecting the progress of work on site to ascertain that a construction project is proceeding in accordance with our designs and meeting our clients' requirements.

Our interaction with the environment and the unexpected circumstances it might present is one of the most challenging, exciting and enjoyable aspects of being an architect. However, it is also when we are most likely to face potentially hazardous situations that pose a risk to our safety, health and wellbeing.

When we talk about visiting site, we tend to envisage visiting managed demolition and/or construction sites. Whilst these pose particular risks, arguably it is the vacant sites you may visit before construction has commenced – or possibly even before a project has been conceived – and which are not under active daily management, that present greater risks.

> For the purpose of this guide, we refer to **occupied sites** – those that are occupied by the owner or tenants and are actively managed, including construction sites; and **unoccupied sites** – those that are vacant or disused and not actively managed. In either case the person responsible for the control of the site or their representative may or may not be present during your site visit.

Naturally, returning safely from any site visit is always our intention. Ensuring this happens requires common sense, responsible site behaviour and being ready to respond to hazards, if and when they arise.

In this chapter, we consider six aspects of how to prepare for every site visit you intend to undertake – whether to an unoccupied or occupied site – to ensure you are adequately equipped to conduct your visit safely:

Do not proceed with a visit to an unoccupied site unless you can ascertain that it is safe to do so.

1.1 Site surveys and research
1.2 Planning work
1.3 Site occupation and vacant sites
1.4 Clothing, equipment and personal protective equipment (PPE)
1.5 Weather conditions
1.6 First aid

1.1 Site surveys and research

Gathering site information before your visit, based on formal surveys or previous experience (your own, that of your colleagues or of the site owner), is invaluable. Familiarise yourself with all available information regarding the site before your visit.

Request copies of all site information in the possession of the site owner. This may include site plans, site photographs, footage from digital drone surveys, condition survey reports and details regarding the presence and nature of any known or suspected site contamination, in particular any asbestos-containing materials (ACMs).

Consult any historic maps, satellite images and/or survey information to identify ancient structures and landscape features that may be present on site, bearing in mind once you are on site that such structures and features may be unstable due to decomposition, weathering or vandalism.

If you are aware of existing buildings or structures on the site that you are due to visit, obtain as much up-to-date survey information regarding their condition prior to your visit as is practicable. Identify any confined spaces or unsafe structures that you will need to avoid. This is especially important for vacant, disused, derelict or semi-derelict structures where there may be the risk of fragile floors, stairs or roofs, or where demolition has already taken place and there may be risk from partially concealed basements, the presence of asbestos or live unknown services. If you need to inspect or gather information from these areas, consider how you may do this without putting your safety at risk, for example, by utilising drones and digital technology.

Consult all available site information prior to your visit and verify its accuracy once you are on site.

If you are visiting a building that was constructed or refurbished before 2000, request a copy of the site owner's asbestos management survey to establish whether there is any risk of exposure to asbestos during your visit. The site owner has a legal duty to determine if an asbestos survey is needed.[1] If asbestos is present on site, the survey will record what it is, where it is, how much there is and the condition it is in. If asbestos has been identified on site, make sure you are familiar with its location and condition. Check the survey for details of any caveats or limitations regarding its use, including any areas of the site that may not have been surveyed. Never visit a site where there is a risk of exposure to airborne fibres that are released when asbestos is damaged or disturbed.

All employers have a duty to provide adequate asbestos awareness training to anybody visiting site. Ensure that you undertake this training before your site visit so that you understand how to avoid the risk of exposure to asbestos. Risks associated with exposure to asbestos are covered in more detail in Chapter 3.

Verify the accuracy and currency of all site information you receive from others and bear in mind that conditions may have changed since the site was last visited or surveyed. Consider the age of any record information – in particular any as-built plans – and reflect on whether there may have been subsequent alterations on site since the records were produced.

If you have any concerns regarding the quality or accuracy of site information at your disposal, speak to the site owner or person in control of the site and identify what further information you require for consideration prior to your visit.

Establish whether there are any live services or mechanical plant that may pose a risk on site and whether such services/plant have been safely decommissioned and/or disconnected.

Once you are aware of and understand the potential hazards that may be present on site, prepare an action plan to determine how you will respond to hazards, should they arise. Make sure this is agreed and understood by any colleagues who intend to accompany you on your visit.

Familiarise yourself with details of any asbestos identified on site and do not enter any areas that pose a risk to your health.

Do not approach of attempt to use any services or mechanical plant that is in poor repair.

DANGER OF DEATH

Isolate supply before opening door

Your action plan should be site-specific, recorded in writing, reviewed and, if appropriate, updated prior to every site visit. Provide copies of the action plan to others who you know will be visiting the site, for example, survey companies that you may have instructed on behalf of your client.

Ensure that you are aware of any regulatory restrictions or duties that the site owner or you, as a site visitor, need to be knowledgeable about. This includes any legal duties you may have as an employer with respect to safeguarding the welfare of your employees (explored in more detail in Chapters 3 and 4).

1.2 Planning work

Before you go on site, determine and appropriately plan the purpose of your visit. Ensure you have adequate time to undertake the tasks you propose to embark on without rushing, bearing in mind how site conditions may compromise how efficiently you may be able to work. This is particularly important during the winter when inclement weather or limited daylight hours are more likely to restrict how long you can work safely on site. Avoid the temptation to continue your visit beyond the time it is safe to do so in an effort to get your planned task completed, even if that task takes you longer than anticipated.

Pre-plan the activities you intend to undertake to ensure you have access to any survey equipment that you require once you are on site. This might include arranging for appropriate temporary working platforms, lighting and/or power to be installed before you arrive.

If you require any destructive opening-up to be undertaken on site to complete any intrusive survey work, wherever possible arrange for the opening-up works to be carried out prior to your visit. If the site owner's asbestos management survey has identified the presence of asbestos on site, ensure that a refurbishment/demolition survey is completed prior to any destructive opening-up work. If asbestos is identified, confirm with the site owner that it has been removed by a licensed contractor and that the area you are visiting has been certified fit for reoccupation prior to your visit.

Pre-plan your activities to ensure appropriate access is available for your visit.

Ensure a qualified operative is available to assist you with any specialist access equipment required during your visit.

If you require access to be provided via mobile plant, such as a mobile elevating work platform (MEWP), ensure that you are accompanied by an operative qualified to assist you in the use of such plant and that the terrain is suitable to ensure the plant can be used safely.

Check whether you need any specific training, or a permit to work, to enter the areas of the site you intend to visit, as these may be required for high-risk activities. For example, work in a confined space such as a culvert or sewer will be subject to management controls.

1.3 Site occupation and vacant sites

Ascertain from the site owner whether the site is occupied and, if so, whether the occupants need to be informed of, or consent to, your visit.

Consider if children or vulnerable young adults may be present on the site you are visiting. If they are, it is worth checking whether any consents, permissions, or DBS (Disclosure and Barring Service) checks are required prior to attending and if the client has a Safeguarding Policy for you to adhere to.[2] If you are visiting an occupied healthcare facility, find out what infection control procedures are in place and make sure you comply with them.

If the site is occupied, determine who is responsible for control of the site and whether there are any specific instructions that you need to be aware of before your visit, including details of any managed emergency evacuation procedures.

Always consider the risk of illegal occupation of vacant sites. Squatters may have caused damage intended to compromise safe access/egress or may have taken measures to prevent eviction that make it unsafe for you to visit. If you suspect this to be the case, avoid entering the site and inform the site owner. Illegally occupied sites may also pose a risk with respect to past or present illegal or illicit activity. For example, toxic

Make sure you comply with any infection control procedures in place at the site you visit.

residues from drugs manufacturing or discarded needles from intravenous drug use pose a risk of poisoning, infection and/or injury.

1.4 Clothing, equipment and personal protective equipment (PPE)

Before leaving for site, enquire about the general site conditions and check the weather forecast to ensure you are dressed appropriately. Ensure you wear sensible and well-fitting footwear and clothing.

Avoid loose-fitting clothes that may snag or get caught on overgrown vegetation or projections.

Wear full-length sleeves and trousers to minimise exposed areas of skin that will be vulnerable to cuts, scrapes and infection. Your footwear should be appropriate for the site conditions. If you are visiting an unoccupied or construction site where there is a risk of debris on the floor that could otherwise injure, you must always wear site boots with a steel soleplate and toecap.

Ensure you are dressed appropriately for the site you are visiting and the activities you plan to undertake whilst there.

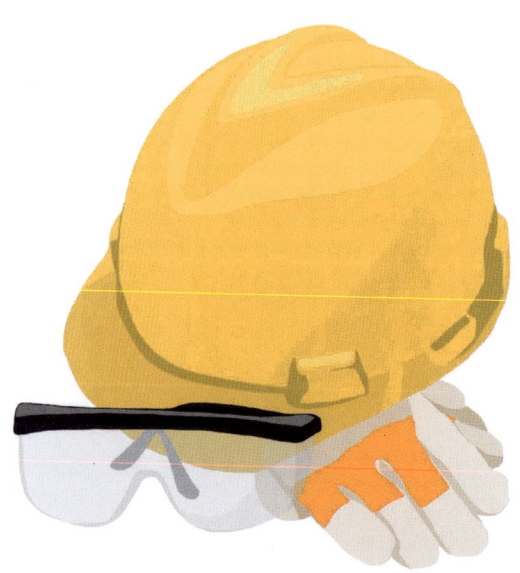

PPE you may require for a site visit includes a hard hat, safety gloves and safety goggles.

Consider whether you need to wear protective gloves and/or any head and eye protection, particularly on semi-derelict sites where there might be a risk of dust, contamination or loose falling debris, or on exposed sites where you may be working outside for an extended period.

Remember that overexposure to hot, sunny weather can be as much of a risk to your safety and wellbeing as cold or inclement weather.

If you are visiting an occupied site, check beforehand if there are any requirements for personal protective equipment (PPE) or specialist clothing and whether this will be provided for you. If you are required to wear specialist protective clothing that is supplied for you, check that it is in good condition, that it fits you properly and that you are wearing it correctly before commencing work on site. Do not use PPE that appears to have been damaged, altered, tampered with or repaired.

Unless you are in a designated safe area (for example, the site office), the PPE you should wear at all times when visiting a demolition or construction site should include:

Ensure any PPE you require is in good condition and fits you properly.

- safety rated site boots/shoes, with ankle support and a protective midsole and toecap[3]
- a hard hat
- hi-vis clothing
- safety gloves
- eye or hearing protection, if appropriate.

If you are in any doubt about what PPE is required, seek the advice of the person responsible for control of the site before entering any active work areas.

Your employer has a legal duty to provide you with the PPE that you require to undertake your site visit safely and you should not commence your visit without it. Often a contractor will have a supply of PPE that you may use, but this may not always be the case or the PPE available may not be an appropriate size or fit.

Check with the person responsible for control of the site prior to your visit exactly what PPE will be available to you so you know whether your employer will need to provide you with the necessary PPE instead.

Check that you are wearing any PPE correctly. Ensure your hard hat is fitted square on your head, is not loose and does not obstruct your vision.

Ensure you are wearing any PPE correctly. Your hard hat should be fitted square on your head.

If hazardous activities cannot be postponed during your visit, ensure you have the correct PPE for protection.

If you drop your hard hat from height on to a hard surface, notify the person responsible for control of the site and replace it.

Some PPE may be task-specific, for example, different safety gloves protect against different types of hazards. If you are unsure whether the PPE you have is suitable, confirm with the person responsible for control of the site prior to proceeding.

If you observe a particular hazard on site for which you are not adequately equipped and it is not possible to avoid the hazard, return to the site office to obtain the correct PPE. For example, if you are required to visit an area of the site where site personnel are cutting materials and there is a risk of flying debris, if the work cannot be stopped briefly to allow you safe access, ensure you are equipped with impact-rated goggles.

Before leaving for site, make sure you have a fully charged mobile telephone and that an appropriate person has the telephone number so they can contact you in the event of an emergency (see Chapter 2 for more detail). Keep your mobile telephone switched on for the duration of your visit with the ringtone volume set to an audible level to ensure that you can hear it. Where the mobile telephone signal strength may be unreliable, identify areas on or close to the site where a sufficiently strong signal is available in case you need to make an emergency call.

Depending on the time of year and conditions on site, consider whether you need to take a torch with you. If you do, check that it is in good working order and either use a wind-up torch or ensure that the batteries are fully charged.

When visiting a remote site, take a bottle of water with you and, if necessary, food and any other provisions.

1.5 Weather conditions

It is important that you check the weather forecast and consider recent weather conditions before going on site. Bear in mind that peak flows in streams and rivers may not occur until several hours and possibly days after heavy rain and can cause unexpected flooding during otherwise benign weather conditions. Inclement weather will also create hazardous working conditions on open or derelict sites and may increase the risk of subsidence and collapse of unstable structures and landscape features.

1.6 First aid

Cover any cuts, abrasions or other breaks in the skin with waterproof dressings before you go to the site. Equip yourself with a basic first aid kit, which may include guidance on first aid, individually wrapped sterile plasters, eye pads and triangular bandages, safety pins, unmedicated wound dressings, disposable gloves, hand sanitiser and anti-bacterial wipes.

If you suffer from any cuts or abrasions whilst on site, cover them with a waterproof dressing as soon as you can and ensure that you wash, disinfect and re-dress them at the earliest opportunity once you have returned from your site visit.

If you have any medical conditions, such as diabetes or severe allergies, consider whether you need to take any medicine with you in case you are on site longer than anticipated. Judge whether any personal conditions may impact or limit your ability to inspect the site and adapt your work plan to suit. If you are claustrophobic or suffer vertigo, ensure you advise those accompanying you on your visit and avoid situations that may cause you unnecessary stress or distraction.

Equip yourself with a basic first aid kit.

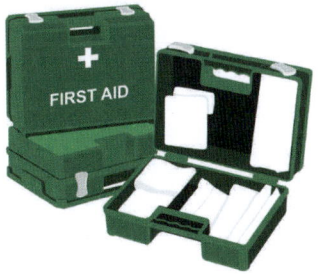

CHAPTER 2:
UNDERTAKING SITE VISITS

When you visit any site, you have a responsibility to be mindful of your own safety and that of others who may be accompanying you on your visit or might otherwise be in attendance whilst you are there. Irrespective of whether you have never visited a site before or have many years' experience you always need to be alert to unexpected or unanticipated hazards, so you are ready to act appropriately.

Never be complacent or take your own safety for granted. If you do, you are more likely to make inappropriate decisions that may put you and your colleagues in danger, including those that you may need to call upon to help you in the event of an accident or emergency.

Never take unnecessary risks. If you are in any doubt about whether it is safe to proceed or continue with your site visit, do not do so. If a situation feels unsafe, it probably is, and you should remove yourself from it.

If you are an employer, or are otherwise responsible for supervising a colleague, you have a duty to ensure that those working under your supervision are not exposed to potentially hazardous situations. Make sure employees and colleagues understand the need and feel confident to raise concerns if they come under pressure or feel they are expected to take risks whilst on site, possibly to satisfy a client or contractor or to fit in on sites where the culture around health, safety and wellbeing might fall short of good practice or may be in breach of health and safety legislation.

Never visit a site if you are under the influence of alcohol or drugs, including prescription medication that may impair your cognitive functions.

In this chapter, we consider 12 aspects regarding how you should conduct your site visit, whether to an unoccupied or occupied site, to ensure you are able to undertake your visit safely:

2.1   Lone working
2.2   Personal site safety
2.3   Person responsible for control of the site
2.4   Induction and orientation

2.5   Safety signage
2.6   Navigating around the site
2.7   Site vehicles and mobile plant
2.8   Inspecting construction work
2.9   Communication with site personnel
2.10 Site behaviour
2.11 Action in the event of an emergency
2.12 Post site visit activity

Avoid visiting sites alone, particularly remote, vacant or unoccupied sites.

2.1 Lone working

Whenever possible, avoid visiting sites alone, particularly remote, vacant, or unoccupied sites. If lone working is unavoidable, consider the following precautions prior to undertaking your visit:

- Always identify someone who is willing and able to be your emergency point of contact during your site visit, ensuring that they will be contactable for the duration of your visit.
- Provide your emergency contact with details of your site visit, including its address, your mobile telephone number, when you anticipate arriving and how long you expect to be on site.
- Follow this up by confirming when you have arrived on site, any change to your anticipated departure time and then again to confirm when you have left site.

2.2 Personal site safety

When you arrive on site, carry out an initial assessment of the site conditions to confirm whether there are any significant hazards or unexpected site conditions that you did not identify from your pre-visit site research, and which may have an impact on your proposed plan of work.

Identify the main site access routes and entrance and exit points, particularly those in the vicinity of the area(s) in which you will be working,

Remain vigilant to any change in circumstances that may compromise your access to and from site, such as livestock in the vicinity.

so you know where to go in the event of an emergency. Remain vigilant to any change in circumstances that may compromise your access to and from site. For example, if you are working in the vicinity of any livestock, particularly cattle, regularly check that you are not disturbing them and that they do not block your site access.

2.3 Person responsible for control of the site

Prior to your visit, ascertain whether the site is unoccupied or occupied and establish who is responsible for control of the site. Ensure you have their permission to access all the areas of the site you intend to visit, including details of any security arrangements, particularly for unoccupied sites. Exchange contact details and emergency contact telephone numbers with the person responsible for control of the site. For demolition and construction sites this will be the principal contractor's site manager or site agent.

Agree the purpose of your visit with the person responsible for control of the site, as well as your anticipated plan of work and what requirements/provisions need to be in place to enable you to carry out your work safely. Check whether you need to undertake any specialist training or certification before going on site. For example, if you are working adjacent to live transport infrastructure you may require specific training and certification to ensure you are aware of the particular hazards and precautions that this work may involve.

The person responsible for control of the site is legally responsible for all health and safety matters on site and should be your first point of contact for all queries and concerns whilst you are on site.

2.4 Induction and orientation

When you arrive on an occupied site, identify and present yourself to the person responsible for control of the site and sign in.

If it is your first visit to the site, ensure you receive a site induction. This will identity site-specific hazards and the site safety and emergency procedures you are required to follow, the use and type of PPE, location of welfare facilities and where there is a first aid station, if required.

If you have visited the site previously, confirm with the person responsible for control of the site the current status of any site works and whether there have been any changes to the induction information with which you need to familiarise yourself. Bear in mind that the site hazards that you need to be aware of during each visit may change as site work progresses.

Re-confirm with the person responsible for control of the site your proposed plan of work and the areas of the site you intend to visit. Ensure you are aware of any areas to which your access is restricted or prohibited. Do not enter these areas unless you have express permission to do so.

Ensure you are aware of and understand the implications of any hazards you are likely to encounter on your site visit.

A site that is tidy and well-organised is likely to be well-run and safe.

Familiarise yourself with all site safety information when you arrive on site.

As a rule of thumb, a site that is tidy and well-organised is likely to be well-run and safe.

If you have any concerns regarding the site management and safety of a site, do not proceed with your visit. Raise any concerns with the person responsible for control of the site and/or the site owner, as appropriate, and request that your concerns are addressed before you re-visit the site.

2.5 Safety signage

A well-managed site will have relevant site safety notices and information updated on a daily basis available for inspection on notice boards adjacent to the site entrance, site office, welfare facilities and in the immediate vicinity of the work being carried out.

CDM SITE SAFETY NOTICE BOARD

Hard hat must be worn

High visibility vest must be worn

Protective footwear must be worn

Eye protection must be worn

First aid
Your first aiders are:

The nearest first aid box is situated:

Accident book

Fire action
1 Sound the alarm by operating the nearest fire alarm call point
2 Call to call the fire brigade
3 Leave the building by the nearest available exit
4 Report to person in charge of assembly point at:

Fire Escape Plan

SE POLICY/INSURANCE

RISK ASSESSMENTS/ METHOD STATEMENTS

REGISTERS/CHECKLISTS

Familiarise yourself with current site safety information at the outset of each visit and as you navigate around site.

Larger sites may have safety information, including current work activities, located local to areas of the site, for example, at the entrance to each storey. Familiarise yourself with the relevant information at the outset of each visit and as you navigate around the site.

Before going on site, you need to understand the different types of safety signage that you are likely to encounter:

- **Circular red and white signs with a diagonal line are prohibition signs**. These mean you must not do something. For example, do not use your mobile telephone.
- **Circular blue and white signs are mandatory signs**. These mean that you must do something. For example, wear safety gloves or eye protection.
- **Triangular yellow and black signs are warning signs**. These alert you to hazards or danger. For example, warning you a substance or contents are harmful or flammable or that industrial vehicles are in operation.
- **Rectilinear green and white signs are safety signs**. These provide you with safety information. For example, identifying an emergency assembly point, fire exit or first aid station.

See Appendix I for more examples of site safety signs.

2.6 Navigating around the site

Remain vigilant and avoid distractions, and distracting others, whilst you are on site, particularly when you are moving around the site. If you need to make or take a telephone call, make sure you find a safe place to do so and remain stationary until you have finished your conversation, unless you are on a designated safe pedestrian route. Likewise, do not walk around when you are referring to, or making, any notes or survey information, including sketching, taking photographs or filming. Do not step or walk backwards.

Avoid carrying heavy, bulky, or cumbersome equipment unaided whilst moving around site as this increases your risk of tripping or falling.

Circular red and white signs with a diagonal line are prohibition signs.

Circular blue and white signs are mandatory signs.

Triangular yellow and black signs are warning signs.

Rectilinear green and white signs are safety signs.

Do not walk around when you are referring to, or making, any notes or survey information.

Remain vigilant at all times, observing site conditions above, beside and in front of you. Do not limit your attention to the floor or areas immediately in front of you.

Do not hurry your visit or rush around. Take your time and try to anticipate or react in a timely fashion to any potential hazards if they occur. Do not proceed and do not take unnecessary risks if in doubt. It is safer to return to site when you have more time or when site conditions improve than to rush to finish your work, risking an accident.

2.7 Site vehicles and mobile plant

When you arrive on site, familiarise yourself with the safe pedestrian routes around the site. Bear in mind that these may have changed since your last visit. On a well-managed site, vehicular and pedestrian routes will be segregated by physical barriers. Keep to the designated pedestrian routes at all times and do not attempt to remove, move or circumnavigate any route barriers.

Whilst walking around the site, avoid any routes that take you through vehicle compounds or in close proximity to mobile plant that is in use. Mobile plant that is reversing poses the greatest hazard to pedestrians on construction sites. If a vehicle marshal is present, always follow their instructions and heed their warnings. If you are asked to assist marshal mobile plant operations, always politely refuse and avoid the temptation to be helpful. Only trained operatives should marshal mobile plant operations.

If you see a mobile crane lifting a load that is about to hit something, warn the person supervising the lift. If you think a load is about to fall from a moving forklift truck, keep clear but try to warn the driver and others in the area. Never walk underneath a raised load.

Never ride in or on mobile plant unless the plant is designed to carry passengers.

Always follow a vehicle marshal's instructions and heed their warnings.

Keep to the dedicated pedestrian routes at all times.

PEDESTRIAN ROUTE

Never walk underneath
a raised load.

If you see mobile plant being driven too fast or operated unsafely, keep out of the way and report the matter to the person responsible for control of the site.

2.8 Inspecting construction work

Prepare in advance for your site inspection, including identifying what you intend to inspect and for what purpose. Speak to the person responsible for control of the site to make sure you will have safe access to the areas you require, that the relevant works are ready for your inspection and what special access equipment you will need access to, if any, including suitable PPE for the areas in which you will be working.

Once on site, keep to your proposed plan of work. Avoid distractions and do not allow yourself to be sidetracked during the course of your inspection. If the contractor requires your advice regarding any design or site queries, make time to address these before or after your planned inspection work.

Make a note of any specific health and safety notices applicable to the areas you are inspecting, the relevant fire alarm points and escape routes. This is particularly important during the early construction stages of a project before the permanent fire detection, protection and separation works are complete.

If site operatives are working in the area that you intend to inspect, speak to the person responsible for control of the site to confirm whether it is safe for you to proceed. If hazardous work is being undertaken, including hot works, noisy works or work that is generating dust, postpone your inspection and re-arrange it for a time when the works will have been completed.

Do not continue your inspection or walk around whilst you make notes or sketches and take photographs. Pay particular attention to observing materials, equipment or incomplete or temporary work that could be a

Speak to the person responsible for control of the site to confirm whether it is safe for you to visit areas where site operatives are working.

48

Permit to work must be obtained

Do not enter any restricted areas unless you have permission and appropriate training to do so.

potential trip hazard or limit your access. Be mindful that dust and site contamination will limit or reduce the slip resistance of floor finishes and inadequately illuminated spaces can pose hazards and should be avoided.

Keep to designated safe routes and do not enter any restricted areas, or areas requiring a permit to work, unless you have received permission and/or appropriate training to do so.

If temporary works, materials or equipment restrict your access, do not attempt to move or adapt them. Speak to the person responsible for control of the site and arrange to have the obstruction removed or arrange to carry out your inspection at another time when it will be safe to do so.

If you require access to work at height, check that any ladder, scaffold, access platform or mobile equipment you use is in good condition. If you are required to utilise a fall restraint system, ensure that it is a fall prevention system rather than a fall arrest system (i.e. the harness and lanyard prevent you from falling from height in the first place rather than arresting your fall mid-drop). Being suspended in a fall arrest harness can result in physical trauma, which can be fatal in as little as 15 minutes.

If a fall arrest system is being used and you cannot avoid its use, make sure in advance that:

- the system has been suitability assessed for your intended use
- you receive adequate training in the use of the proposed system

- the equipment is in good working condition
- there is an adequate rescue plan in place in the event that you do fall from height.

Do not proceed unless you are happy it is safe to do so and do not use specialist access equipment unless you have been specifically trained to do so or are using it under adequate supervision.

2.9 Communication with site personnel

The principal contractor's site manager, or the site agent responsible for control of the site, should be your primary point of contact whilst you are on site. If you witness any unsafe behaviour by site personnel or hazardous conditions during your visit, promptly notify the site manager. Do not instruct site personnel directly unless you need to alert them to immediate danger and it is safe for you to do so. If you are unsure or concerned regarding the work or working practices you have observed during your visit, discuss these with the site manager. Do not wait to raise your concerns regarding any unsafe activity or site conditions because this could result in an accident or accidents that could otherwise have been avoided.

Do not instruct site operatives directly unless you need to alert them to immediate danger.

If you are concerned regarding the safe management of the site by the site manager, notify the principal contractor and/or site owner as soon as it is practical to do so. If your concerns are not addressed, consider whether you should notify the Health and Safety Executive (HSE), particularly if the site poses an imminent risk to the public or to the safety of the site personnel.

Do not respond to queries regarding your design or the works that may be addressed to you directly from site personnel. Ask that all queries are directed via the principal contractor and/or site manager.

Wherever possible, avoid responding to site queries until you have had a chance to properly review and consider the nature of the query, in particular the design risk management implications and potential safety impacts relating to any proposed variations and any knock-on effects to your design.

2.10 Site behaviour

Obey the site signs, rules and instructions at all times. Do not imitate bad, irresponsible, or unsafe behaviour. Do not proceed with your visit if you are in any doubt about whether it is safe to do so and seek advice from the person responsible for control of the site. Do not proceed with your visit if you experience uncooperative and/or coercive behaviour from any site personnel or the person responsible for control of the site.

2.11 Action in the event of an emergency

If you do have an accident or find yourself in a dangerous situation, try to remain calm. If possible, attempt to reach a place of safety away from immediate or imminent danger and then raise the alarm or attempt to contact help as soon as you are able.

Take time to assess your situation and to consider the best course of action. If you are on site with colleagues, stay together unless it would be unsafe to do so. Do not take any unnecessary risks with respect to your own or others' safety, and prioritise your personal safety (i.e. abandon any survey equipment, notes or other similar property).

As soon as it is safe to do so, contact the emergency services and/or your emergency point of contact.

2.12 Post site visit activity

Before you leave the site, check that any access equipment you have used is properly stored and secured to prevent unauthorised use. If you have been authorised to use any site services, and are competent to do so safely, ensure that these are turned off or isolated, as appropriate.

As you leave, check that all access points are adequately secured after you to avoid unauthorised access and notify the person responsible for control of the site that you are leaving. If you have visited a site alone, do not forget to notify your emergency contact that you have left the site.

Ensure that any site services you have used are turned off or isolated before you leave the site.

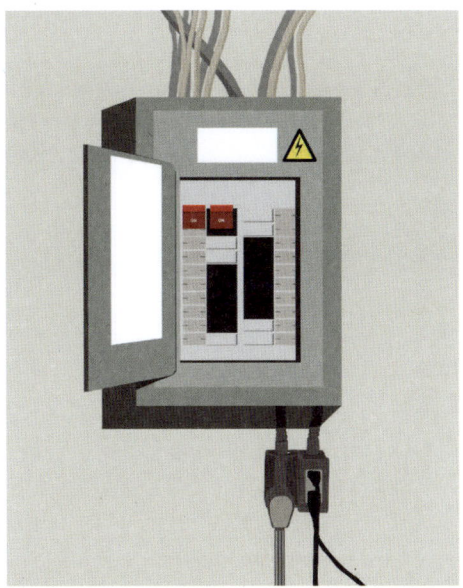

Check that all access points are adequately secured when you leave the site.

Once you get back, ensure you wash your hands before consuming any food or drink. Change out of any soiled clothes and ensure they are washed as soon as possible, and certainly before being used again. Do not store muddy or damp site clothes in cupboards or drawers where they could contaminate other clothing or could decay.

Update any site notes and survey information records to include details of any site features or hazards that you have observed during your site visit that have not otherwise been recorded. Pass on relevant details of any hazards or any concerns you have regarding the site to the site owner and notify any colleagues as appropriate.

Prepare written records of any areas of concern that you noted whilst on site, including photographic evidence where possible.

In particular, make a note of any concerns you may have regarding any health, safety and wellbeing issues and ensure these are shared with the

Ensure any PPE you have used is cleaned and stored appropriately.

PER architects

10 Covent Road
London UK
SW1P 4DJ

tel +44 20 7932 8200
fax +44 20 7932 8201
architects@per.co.uk
www.per.co.uk

Site Inspection Record – Issue 01.07.19

Job No:	10343		Project:	Margate Wood
Notice No:	00	Rev: 01	Contractor:	HCD
Inspection Date:	25.06.19		Issued to:	HCD
Notice Date:			Issued by:	EPR Architects

14 Location: No.1 all floors	25.06.19
Observation: Door handle specification	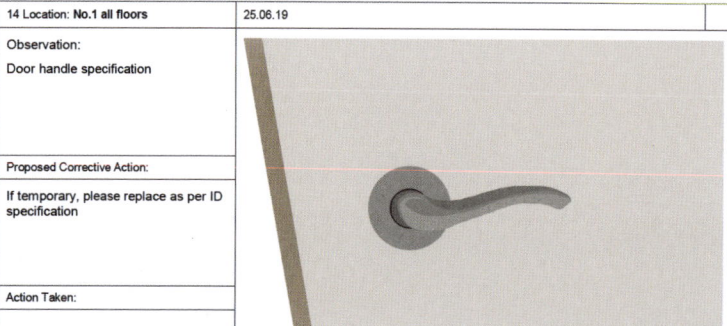
Proposed Corrective Action: If temporary, please replace as per ID specification	
Action Taken:	

6 Location: No.1 All floors	25.06.19
Observation: Cork exposed.	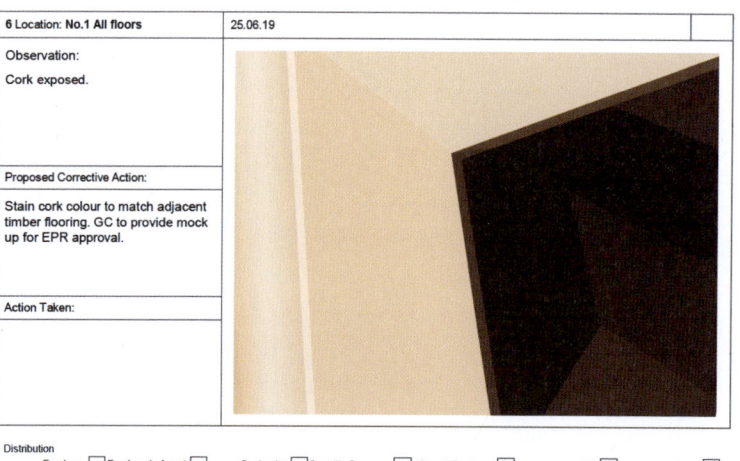
Proposed Corrective Action: Stain cork colour to match adjacent timber flooring. GC to provide mock up for EPR approval.	
Action Taken:	

Distribution

Employer [x] Employer's Agent [x]   Contractor [x] Quantity Surveyor [ ] Struct'l Engineer [x]   File [x]   Other [x]

Distribution

Employer [x] Employer's Agent [x]   Contractor [x] Quantity Surveyor [ ] Struct'l Engineer [x]   File [x]   Other [x]

Prepare written records of your site visit, including photographic details where possible.

site owner, contractor (including the principal contractor) and project team, as appropriate.

Make a written record of any queries raised by the contractor whilst you were on site. If you provided a response whilst still on site, review the advice you gave and confirm details in writing, particularly noting any change in your advice that may be necessary once you have reviewed your position.

If any variations to the project design are required as a consequence of your visit, carry out a design risk management review of the proposed variation. Take into account whether the variation will have any detrimental impact on safety strategies adopted more generally for the design of the project as part of your design risk management strategy, which are considered in more detail in Chapter 4.

CHAPTER 3:
SITE
HAZARDS

Every site is unique. It can be:

- a remote rural location
- a greenfield site
- a brownfield site
- a disused industrial unit
- an unoccupied or occupied office building
- an ancient castle or ruin
- a demolition or construction site.

Each will pose unique hazards that you need to consider and manage to ensure you are able to complete your visit safely and return home without incident.

Identifying and providing guidance regarding every possible hazard you may face when visiting site is beyond the scope of this guide.

In this chapter, we consider 15 key aspects regarding the more significant or typical hazards you might face during a site visit and provide guidance on how you can consider managing these to ensure you remain safe:

3.1   Site assessment
3.2   Site contamination
3.3   Falls from height
3.4   Slips and trips
3.5   Unsafe structures
3.6   Excavations
3.7   Enclosed spaces
3.8   Confined spaces
3.9   Respiratory hazards (dust and fumes)
3.10 Noise
3.11 Hazardous substances
3.12 Fire safety
3.13 Manual handling
3.14 Geological, man-made landscape or hydrological features
3.15 Flora and fauna

Each site is unique
and will pose unique
hazards.

3.1 Site assessment

When you arrive on site, carry out an initial assessment of the site conditions to confirm whether there are any significant hazards or unexpected site conditions that you did not identify whilst preparing for your visit, and which may have an impact on your proposed plan of work.

Start by identifying the main site access routes and entrance and exit points, particularly those in the vicinity of the area(s) in which you will be working, so you know where to go in the event of any emergency. Check that the quality of your mobile telephone reception is adequate. Remain vigilant to any change in circumstances that may compromise your access to, from and around the site.

3.2 Site contamination

If you are visiting a building constructed or refurbished before 2000, you need to be particularly mindful of the potential presence of asbestos, which still kills around 5,000 workers each year, more than the number of people killed in road traffic collisions.[1] If you inhale asbestos fibres, which are released into the air when asbestos is disturbed or damaged, you could suffer from fatal and serious lung diseases, including mesothelioma – asbestos-related lung cancer – asbestosis and pleural thickening. There are no immediate symptoms associated with exposure to asbestos and symptoms can take up to 40 years to appear. Once diagnosed, it is often too late to do anything about it. Therefore, it is important to protect yourself during your site visit.

Request and familiarise yourself with the site owner's asbestos register and/or asbestos refurbishment and demolition survey before you visit the site. This will provide you with information on the location, amount and condition of any known or presumed asbestos, particularly any areas where loose asbestos dust or fibres have been identified and to which access should be prohibited. Do not enter prohibited areas unless the site owner can provide you with up-to-date documentary evidence that it is

Familiarise yourself with details of any site contamination before your visit.

safe to do so, for example, certification provided by a competent person confirming the area is safe for reoccupation following the removal of the asbestos. If you suspect there is asbestos present on site, particularly loose asbestos fibres, dust or debris, vacate the area immediately, ensuring anyone else in the affected area does likewise and notify the owner of the site. Also bear in mind that asbestos fibres, dust and debris can be carried on your shoes and clothing and could pose a contamination risk outside of the site, for example, on public transport, back in your studio or at home, so take care to minimise any exposure by thoroughly cleaning your shoes and clothing as soon as it is possible to do so.

Asbestos encapsulated in products/materials that retain their integrity does not pose a risk if left undisturbed. The HSE must be notified 14 days prior to undertaking any licensed asbestos removal.

Examples of asbestos-containing materials that could pose a hazard include asbestos insulation boards (often found in fire doors, firebreaks, partitions and fireplaces), corrugated roofing panels, pipe lagging and gaskets for electrical or mechanical services installations, sprayed structural or fire-retardant coatings on steelwork.

Similarly, if you suspect the site is contaminated with any other substances that may be harmful to your health, vacate the area and notify the owner to arrange for sampling and contamination testing. Site contamination can be a particular risk on brownfield, industrial and mining/mineral extraction sites prior to remediation or where slurry or slag materials may have been discarded. Old pipework and paint finishes may contain lead which can be harmful if ingested.

3.3 Falls from height

Over the last five years, falls from height have been responsible for 51% of fatal injuries to workers on construction sites and account for 33% of all main accidents over the last three years.[2]

Ensure any ladder you
use is properly secured
and in good repair
before use.

If you need to inspect work at height, it is likely you will need to make use of a scaffold, mobile tower scaffold, temporary working platform, or mobile elevating work platform (MEWP). Always ensure that any scaffold or working platform is safe before you access it and check the date on the inspection tag which should be within the past week. If you have any doubts, speak to the site manager who will be able to confirm when the scaffold was last inspected and certified as safe for use.

If you need to use a ladder to access a scaffold or working platform, ensure that it is secured at the top so that it will not slip and that it extends five rungs or 1m above the stepping-off point (if there is no other alternative firm handhold). A leaning ladder should be secured at an angle of approximately 75 degrees. Always maintain at least three points of contact with the ladder as you ascend and descend it, for example both hands and a foothold, or both feet and a handhold.

Never use a ladder with broken rungs, which is not securely fixed, or which is relying on wedges at the bottom or another person to keep it secure.

Once you reach the working platform, ensure that there is adequate edge protection to prevent you falling off. This should include a toe-board, a handrail at a minimum height of 950mm and an intermediate guardrail at 470mm centres.

If you need a mobile tower scaffold to carry out an inspection, confirm with the person responsible for control of the site that:

- it has been erected by a trained, competent and authorised person
- the mobile tower is within the maximum height specified by the manufacturer
- you have been briefed on the safe use of the scaffold.

Once you have gained access to the working platform of the mobile tower, make sure that the access hatch is closed behind you to prevent you, or anyone accompanying you, from falling.

Ensure any working platform you access has adequate edge protection.

Ensure you are adequately briefed prior to using any mobile tower scaffold.

Check the number of people accompanying you on a MEWP is within the safe working load.

If a scaffold obstructs your inspection, never attempt to modify it yourself. Speak to the person responsible for control of the site who will arrange for a competent person to carry out the necessary modifications.

If you are provided access via a MEWP, confirm with the person responsible for control of the site beforehand whether the ground load-bearing capacity has been assessed and that the MEWP is suitable for the terrain in the area of proposed use. This is particularly important on soft ground or where ground may have been disturbed by ground works.

The number of people accompanying you in the MEWP should be within the safe working load, which you should check on the information plate fixed to the MEWP.

Make sure you are provided with an appropriate harness and lanyard before getting on to the MEWP and then attach your lanyard to the designated anchorage point inside the platform before it is elevated. The only exception to this is if you are working over, or near to, deep water to ensure you are not trapped in the MEWP underwater in the event of an accident.

If you need to access a flat roof area, check that it is not a fragile roof or a non-fragile roof with fragile areas (for example, unprotected rooflights, which may be covered in dirt, algae or moss and not clearly visible, or may not have been designed to be walked on) or materials that may become

Make sure you are
provided with a harness
and lanyard before
getting on to a MEWP,
and that it is connected
to the anchorage point
before elevation.

Check all fragile areas of roof are adequately protected before you access the roof.

fragile over time that you might fall through. Corrugated asbestos cement roofs are a typical example of a fragile roof that can be a hazard. You should also check that the roof has suitable edge protection and that all voids, holes or fragile panels (including roof lights) are protected with secure, load-bearing covers.

Avoid working near to unprotected edges where you may fall, including inadequately covered wells, open manholes or excavations, maintenance shafts, and structural openings such as lift openings and riser shafts.

3.4 Slips and trips

Slips, trips or falls on the same level account for 31% of work-related injuries in the construction sector.[3] Poorly maintained or unmanaged sites pose a risk of an increased likelihood of slips or trips. The presence of contamination by liquids or dust and general site debris increases the risk of slips. Inadequate lighting or uneven floor finishes increase the risk of trips. Exercise caution as you navigate around site, remaining vigilant to any potential slip or trip hazards.

3.5 Unsafe structures

Remain vigilant during your visit to the stability and condition of any buildings or structures you may need to enter or work close to. Avoid unsafe or fragile structures, for example, floors, stairs and roofs that appear to be in poor condition and any overhead structures through which loose debris may fall.

If you have concerns about the structural integrity of any areas to which you require access, consider alternative means of access for example, requesting the installation of a temporary working platform or employing drones for visual inspections, remembering to check beforehand if there are any licensing or legal restrictions on drone use.[4] If there are no alternative means of safe access, you should request structural advice is provided before you access those areas.

Check there is adequate light, ventilation and edge protection for the areas of the site you intend to visit.

Do not enter areas with limited access or areas where you cannot readily determine that the conditions are safe to do so.

Check that there is sufficient light, ventilation and edge protection. Avoid restricted spaces and areas where there is a risk you may be trapped. Check that any doors or gates providing access to the areas where you need to work can be opened from the side you are working on to avoid being locked in.

3.6 Excavations

As a general precaution, avoid entering any deep excavations, as these can pose a particular risk of falls from height or burial if they are unstable or are inadequately supported. Consider whether the information you require could be obtained remotely or by alternative means. Do not approach unfenced excavations.

Check that the sides of any excavation you are required to inspect are adequately supported.

If it is necessary for you enter a deep excavation, ask the person in control of the site to carry out an appropriate gas detection test beforehand to ensure that there is no gas present. If you feel unwell whilst inspecting an excavation, leave the area immediately and notify the person in control of the site.

Before you enter an excavation, check that the sides of any excavation you may be required to inspect are battered back, stepped or shored if there is a risk of the sides falling in, regardless of depth.

The site manager should prevent site vehicles from operating in the vicinity of, or approaching, any open excavations to avoid the risk of damage or collapse. If you see vehicles approaching an excavation where there are no stop blocks or similar measures in place to prevent vehicles getting close to the edge of the excavation, immediately warn anyone working in the excavation and report the matter to the site manager.

Leave the excavation immediately if there is any sign of movement and advise anyone else working in the excavation to do likewise. Only access the excavation via a fixed staircase or ladder; do not climb the sides of the excavation or use the excavation props to enter or exit the excavation.

Do not enter deep excavations if you are working alone.

Check that vehicles are prevented from operating in the vicinity of any open excavation you are required to inspect.

3.7 Enclosed spaces

Cramped spaces pose a risk due to their size or the difficulty posed by working within them. This can result in musculoskeletal injuries or problems with evacuation if you have a fall or injury. A typical example is a small roof or loft space.

Avoid entry to any enclosed spaces unless you are certain it is safe to do so, and you have adequate arrangements in place to deal with an emergency.

3.8 Confined spaces

A confined space is a space which is substantially (though not always entirely) enclosed and where there is also an identifiable risk to your health and safety, or such risk is reasonably foreseeable (for example, a lack of oxygen or an accumulation of harmful gases or noxious fumes).

Remain vigilant to potential changes in circumstance if you are required to enter a confined space.

Typical examples of a confined space include sewers and storage tanks. However, a confined space is not necessarily enclosed on all sides and may not be small or difficult to work in. Examples include service ducts and risers, enclosed rooms (particularly plant rooms), unventilated or inadequately ventilated rooms and ceiling voids, and deep excavations and trenches. Depending on the nature of a space and the activities taking place within it, the status of the space may change depending on the circumstances. For example, heavy rain may present a foreseeable risk of drowning in a space not usually considered confined.

Some of the potential risks associated with confined spaces include:

- the risk of fire or explosion due to the presence of flammable substances (for example, fumes left in a disused fuel tank or leaks from services that have not been adequately isolated)
- excessive heat
- fumes, vapours and/or oxygen deficiency, which could all lead to asphyxia or unconsciousness, or toxic gas

- the ingress or presence of liquids or solid materials that can flow and could drown or submerge you and prevent you from breathing.

Avoid entry to any confined spaces and never do so if you are working alone. If entering a confined space is unavoidable, you must ensure that you have a safe system of work in place beforehand that includes adequate arrangements to deal with any emergency.[5] If you feel unwell whilst inspecting a confined space, leave the space immediately and notify the person responsible for control of the site. Do not risk your own safety by entering a confined space in an attempt to rescue someone who has collapsed in there, unless you are trained and competent to do so and are acting in accordance with the emergency rescue procedures for the site.

3.9 Respiratory hazards (dust and fumes)

Dust-making activities on sites, particularly demolition sites, pose hazards that potentially have long-term consequences for your health if you do not take adequate precautions to protect yourself. Breathing in hazardous dust and fumes is the biggest cause of long-term health issues in the construction industry. Occupational asthma,[6] silicosis[7] (sometimes linked to lung cancer) and chronic obstructive pulmonary disease (COPD)[8] are a few of the conditions that commonly result from prolonged exposure to dust on construction sites.

Activities such as cutting or grinding concrete blocks or roof tiles can release harmful dust into the air. This is a particular risk if the activity is taking place in an enclosed or poorly ventilated space and without the site personnel employing appropriate measures to minimise dust. Site personnel should use wet cutting (i.e. using water to damp down dust) or power tools with a dust extractor or collector for such work.

Exposure to paints or resins that have high levels of solvents can cause headaches, sickness, drowsiness, poor coordination and dermatitis or skin problems.

If there is a risk you will be exposed to harmful dust or fumes, avoid the area. If this is not possible, arrange for access when work is not being carried out or ask the person responsible for control of the site to equip you with suitable respiratory protection, such as a face mask or respirator.

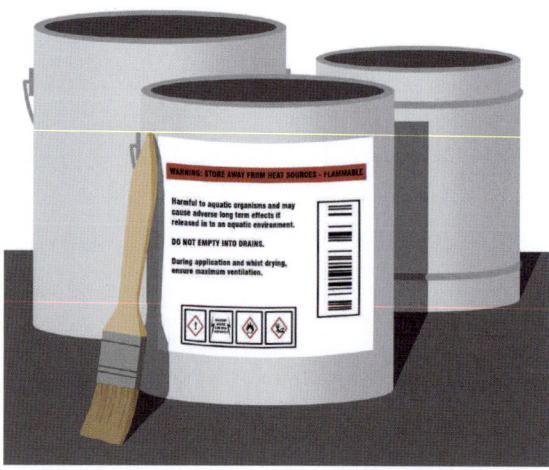

Avoid exposure to paints or resins that have high levels of solvents.

Avoid areas with harmful dust or fumes. If this is not possible, make sure you are equipped with suitable respiratory protection.

Use hearing protection if you need to work in the vicinity of noisy works.

3.10 Noise

Exposure to noise can cause temporary hearing loss and may lead to permanent, irreversible damage. If you need to inspect an area of site where noisy construction activities are taking place that cannot be stopped, ask the site manager for hearing protection, and use it for the whole time you are in the vicinity of the noisy works.

Hearing protection zones are mandatory for all works exceeding 85dB(A). As a rule of thumb, noise levels from construction activities are likely to be hazardous if you have to raise your voice to be heard when someone is standing two or more metres away from you.

3.11 Hazardous substances

Prior to commencing your activities on site, confirm with the person responsible for control of the site whether any hazardous substances may be present in the areas you intend to visit. Substances that are hazardous

Avoid contact with any hazardous materials.

to health include chemicals, products containing chemicals, fumes, dust, vapours, mists, nanotechnology, gases, asphyxiating gases and biological agents, including germs that cause disease.

Under the Control of Substances Hazardous to Health Regulations 2002 (COSHH), the person responsible for control of the site has a duty to carry out an assessment of any hazardous materials and implement control measures to minimise the risk of harm associated with these.[9] Whilst you should not be directly involved in the use of COSHH-assessed substances, the person responsible for control of the site ought to advise you of any control measures which may have an impact on your site activities.

Whilst inspecting work on site, avoid contact with wet cement, screed or concrete because it can cause serious chemical burns to your skin. Avoid enclosed spaces with no visible means of ventilation where toxins or noxious vapours may pose a health hazard.

3.12 Fire safety

If you discover a fire whilst on site, the first thing you should do is raise the alarm either by activating the fire alarm on an occupied site or contacting the emergency services on an unoccupied site.

If you hear a fire alarm whilst on site, head straight to the nearest assembly point and await further instructions from the person responsible for control of the site.

Unless you are – or someone else is – in immediate danger, do not attempt to tackle a fire on site yourself. Different types of site fires will require different fire extinguishers or equipment. Tackling these should only be done by trained operatives.

Familiarise yourself with the location of your nearest fire and assembly points as you navigate around the site.

If you are alerted to a fire, head straight to the nearest assembly point and await further instructions.

The most common fire risks on construction sites are uncontrolled hot works and poor housekeeping, leading to a build-up of waste material. If you see conditions on site that you think pose a fire risk, advise the person responsible for control of the site before you leave the site, or sooner if you think there is an imminent danger of fire.

3.13 Manual handling

Injuries while handling, lifting or carrying on construction sites amount to 25% of the main accidents on construction sites over the last three years.[10] Whilst you are on site, never attempt to lift or move any heavy site materials. If you find your access is restricted by site materials, speak to the person responsible for control of the site about making appropriate arrangements to remove the obstruction.

Alert the person responsible for control of the site if you witness uncontrolled hot works being carried out on site.

3.14 Geological, man-made landscape or hydrological features

Typical examples include redundant or historic excavations, wells, sink holes or mine shafts, including partially filled or concealed basements. Not being aware of your surroundings, or features masked by undergrowth, can lead to life threatening falls.

Deep or fast flowing water can be dangerous, particularly following poor weather conditions and if concealed by overgrown vegetation or adjacent to steep embankments. Here the risk of falling in and not getting out unaided may be greater and could lead to drowning.

If you are exposed to soil or freshwater (such as from a river, canal or lake) and it gets into your mouth, eyes or a cut, you are at risk of contracting leptospirosis (also call Weil's disease).[11] This is an infection you can also

Avoid contact with soil, wildlife and freshwater.

catch from animals, typically from the urine of rats and mice, but also cows, pigs and dogs. Whilst leptospirosis is rare in the UK, there is a higher risk of you being exposed to it where there is evidence of rat infestation and when adjacent to canals, rivers and sewers.

The disease starts with flu-like symptoms such as a headache or muscle pains. More severe cases can lead to meningitis, kidney failure and other serious conditions. In rare cases the disease can be fatal.[12] You can reduce your chances of catching leptospirosis by ensuring you avoid coming in to contact with any dead animals or sources of fresh water whilst on site and by exercising good basic hygiene, particularly avoiding hand-to-mouth or hand-to-eye contact whilst on site and until you have been able to wash your hands.

3.15 Flora and fauna

Take care on unoccupied sites that are poorly managed and where access may be restricted due to overgrown vegetation. Typical risks will include

Avoid contact with
giant hogweed.

stings, rashes and pricks by thorny plants and nettles. More seriously,
giant hogweed, a close relative of cow parsley and an invasive species
that can reach over 3m in height, has sap that can cause severe skin burns
that may result in blistering, boils and scarring. Giant hogweed is widely
distributed in the wild, particularly on sites adjacent to infested woodland,
heathland or common land.[13]

Nesting or breeding birds and animals can pose a danger, if threatened.
Gulls in particular are known to be aggressively territorial and may swoop
towards you or defecate or regurgitate food on you to warn you off. If you
are the target of a swooping gull, the best defence is to raise your arms to
protect your head and then move away. Resist the temptation to wave your
arms around to scare a gull off as this could make it more agitated.

Birds carry a number of diseases that can be harmful to humans, usually
through exposure to contact with infected birds or more commonly
through inhalation of airborne particles from their dried faeces, particularly

Avoid areas with
nesting birds. Use your
arms to protect your
head and vacate the
area if you are attacked
by gulls.

Do not enter enclosed
spaces where large
numbers of birds
may be or have been
nesting.

Do not enter poorly ventilated enclosed or confined spaces.

where large numbers may be or have been nesting. Human infection can result from brief, passing exposure to infected birds or their dried contaminated droppings, and can cause acute respiratory disease.[14]

Other lung diseases, sometimes known as extrinsic allergic alveolitis can develop if you are exposed to microbes, which form mould on vegetable matter in storage. These include mouldy straw, hay or grain, stored in confined spaces such as poorly ventilated buildings where it is likely you could be exposed to inhaling spores and other antigenic material.[15]

Insect stings are a common hazard for which you should be prepared. If you are allergic to insect stings, especially if you are at risk of anaphylaxis if stung, ensure you have appropriate medication with you and that anyone accompanying you on site knows how to administer your medication if you are unable to do so.

As well as stings, you need to be wary of bites from ticks which can cause Lyme disease, a bacterial infection that causes flu-like symptoms and a circular red rash, that can appear up to three months after being bitten and usually lasts for several weeks. Treatment requires a course of antibiotics and in severe cases may require treatment over several months. Whilst only a small number of ticks are infected with the bacteria that cause Lyme disease, they are found all over the UK. High-risk areas include grassy and wooded areas in southern England and the Scottish Highlands, particularly in areas where deer have been grazing.[16]

Caterpillars of the oak processionary moth, a European species that found its way to the UK in 2006, have hairs that can cause an unpleasant rash, asthma attacks and other allergic reactions if they or their nests are touched.

Bites from spiders in the UK are uncommon, but some native spiders – such as the false widow spider – are capable of giving you a nasty bite, which can cause nausea, vomiting, sweating and dizziness. Seek medical help immediately if you have any severe or worrying symptoms after a spider bite.[17]

Snakes will sometimes bite in self-defence if disturbed, typically if you accidentally step on one. Some snakes are venomous and can inject venom containing toxins as they bite; however, adders are the only venomous snakes found in the wild in the UK.

If you are bitten by a snake, try to remain calm and do not panic. Snake bites in the UK are not usually serious and are only very rarely deadly. However, you should keep the part of your body that has been bitten as still as possible, loosen any restrictive clothing and seek immediate medical attention.[18]

Seek medical attention at the earliest opportunity if you are bitten by a tick.

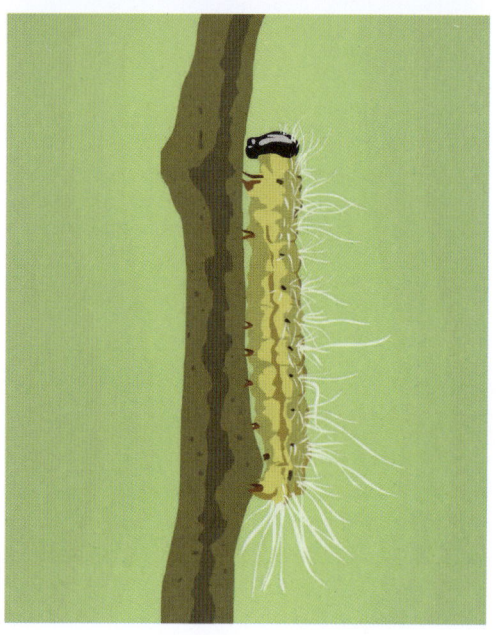

Avoid contact with processionary moths and their nests.

Seek immediate medical attention if you are bitten by a spider and start to exhibit severe symptoms or an allergic reaction.

If you are bitten by a snake, loosen restrictive clothing and seek immediate medical attention.

CHAPTER 4:
DESIGN RISK MANAGEMENT

Everyone in the design team, including your client, has a duty to work collaboratively to ensure that the buildings we design can be constructed, occupied, managed, maintained and refurbished or demolished safely. This is equally important for buildings of all types and scales and applies to work proposed on new and existing buildings, including alterations, extensions, retrofit, renovation, repair and maintenance, change of use or function and other work where relevant.

Architects and lead designers are uniquely placed to influence how effectively safety considerations are embedded into the design process. We have legal, professional and social responsibilities to ensure, so far as is reasonably practicable, that all our projects are well-designed, of good quality, sustainable and safe.[1] We can achieve this by adopting proportionate and integrated approaches to design risk management from the outset of our involvement on every project.

Much of what we do as architects and designers is learnt by example. This may be from precedent studies of projects by other architects, following the example of colleagues, or learning lessons and developing habits based on our own research and experience. However, it is important to avoid repeating the mistakes or shortcomings of past projects or adopting strategies from another project without checking first that it is appropriate to do so. To achieve this, we need to keep abreast of current good practice and design guidance to understand the fire safety, structural safety and public health principles to use as the starting point for each of our projects. This is particularly important, considering the constant evolution of industry legislation and standards that aim to raise the bar on what is proportionate good practice with respect to building safety and design risk management.

Some industry commentators advocate a 'safety at all costs' approach. Our role as architects and lead designers requires us to take a balanced approach to building safety and design risk management in the context of the priorities and limitations of each project. Whilst the safety aspects of your design are important, these should be dealt with proportionately in the context of relevant legislation and your client's brief and aspirations as each project brief and proposed design develops.

In this chapter we consider four key aspects of how you do this by understanding and ensuring your design process employs:

4.1   Principles of design risk management
4.2   General principles of prevention
4.3   Effective communication and coordination

4.1 Principles of design risk management

There is a misconception by some in our profession that risk management is something separate to the design process and that good risk management requires specialist expertise from someone additional to, and possibly remote from, the design team. Whilst you may need to seek specialist advice in connection with complex projects or to help you to address unusual or unique hazards, if you have been appointed as the architect and lead designer *you* ought to have the competence to manage design risks appropriately for the majority of your projects.[2]

The reality is that we are all experts at risk management. Every action that we take in life from getting out of bed to crossing the road involves an element of risk and therefore risk management. We simply undertake these activities so frequently that we tend to do so subconsciously or without forethought unless or until we are faced with an unusual or particularly challenging situation. For example, when you want to cross a road the best way to avoid being involved in a road traffic collision is to avoid crossing the road in the first place. However, this may not be a practical or sensible solution if the only way to reach your destination is to cross the road. So before crossing you will undertake a risk management review during which you will decide where, when and how you cross the road. How you do this will depend on your circumstances. How busy is the road? Do you have clear visibility? Are there managed crossing points close by? How much time do you have? You cannot eliminate the risk of being involved in a collision, but you can reduce the risk to an acceptable or tolerable level. There are no hard and fast rules for determining what this acceptable level of risk is and it is likely this will change depending on your circumstances. For example, crossing the road in one location may be acceptable on a quiet Sunday afternoon but less so, or not at all, on a weekday during rush hour.

Design risk management on your projects should be approached in the same fashion. Constructing, occupying, managing, maintaining, refurbishing and demolishing buildings will always pose hazards to the people undertaking such work and eliminating all of the risks these create is not practical. For example, working at height and cutting masonry materials pose some of the biggest risks to the health, safety and wellbeing of contractors and maintenance operatives but it is not possible, practical or desirable to avoid these activities in all circumstances. The only way to eliminate all risk would be to not undertake a project in the first place.

Your responsibility as a designer is to review all foreseeable risks with your client and the project team and to agree the acceptable, or tolerable, level of risk on your project. As with the example of crossing the road, what is acceptable will be unique to the circumstances of each project, depending on your client's priorities and the nature and complexity of the project.

Foreseeable risks are hazards that you should anticipate, exercising reasonable skill and care, as the result of your design. For example, personal injury as a result of slips and trips are a foreseeable risk of specifying a highly polished floor finish with a high slip potential in an area accessible to the general public.

To decide what is acceptable or tolerable consider:

- each risk versus the benefit of managing the risk
- the environment in which the risk is anticipated
- the foreseeable behaviour of people exposed to the risk.

For example, risks that may not be tolerable in a public space may be acceptable in a controlled environment where access is limited to trained operatives with the appropriate skills, knowledge, experience and behaviours to manage the risk.

Good design risk management benefits from diversity of thought.[3] Avoid tick-box, one-size-fits-all type approaches or solutions, which inhibit creative thinking or innovative design solutions. If you are faced with a situation that you are not sure about, take advantage of the knowledge and experience of others. Treat design risk management as part of the design process and collaborate with the rest of the project team to find the solution most appropriate to the circumstances of your project.

Be careful not to let design risk management or safety issues constrain your design, at either concept, coordination or technical design stage. Exploiting new technology or innovative design can have positive benefits for design risk management. Off-site manufacture, fabrication and assembly are obvious examples of how the construction phase of your project could be better managed or procured to reduce or eliminate some of the hazards associated with on-site construction.

Your duty to do what is reasonably practicable to achieve effective risk management is important. Adopt a proportionate approach to managing risk. When you consider potential hazards, balance the severity and risk of harm in each case against the measures required to control that risk in terms of money, time and trouble. Also balance this with your other project priorities, for example, your client's budget and programme, your aesthetic aspirations – including any desire to innovate or develop unusual design solutions – and any design decisions that could have environmental consequences or might compromise access to your project. You may not be able to avoid or allow for every eventuality, but by working collaboratively with the rest of your project team you should be able to achieve what is practicable to eliminate, mitigate or control risks.

Regularly review your design risk management decisions at each design stage of your project. The opportunities for eliminating risk through the

Explore options for new technology and innovative design to benefit your design risk management.

design will change as the design develops and as your project progresses. Do as much as is reasonable at the time you prepare the design. Do not ignore potential risks at concept design stage, when it is more likely to be reasonably practicable to eliminate them, or you may find you are forced to accept control measures to manage residual risks that compromise the design at the spatial coordination or technical design stage of your project.

When you consider health and safety risks in the context of your other project priorities, it will be inevitable that there will be some hazards and risks that are not reasonably practical for you to eliminate through your design, for example, the need for work at height to clean and maintain your project. In such circumstances, take reasonably practicable steps to reduce or control the risk through your design, employing the principles of prevention and, where relevant, the prescribed principles for the management of building safety risks, so that harm arising from the risk is less likely or the potential consequences are less serious.[4]

Treat design risk management as part of the design process.

Design risk management – the process of deciding which risks you attempt to eliminate, reduce or control – is not always straightforward. Do not rely on a formulaic tick-box approach or duplicate design solutions or strategies taken from previous projects. The context of each of your projects and their associated hazards and risks will be unique. How you respond to this and consider the elimination, reduction or control of risks requires your professional judgement based on your own experience, consultation with your client and advice from the professional team, contractors, specialist suppliers and manufacturers involved in your project.

The imperative to eliminate a risk depends on how significant a risk it is in the context of your project. The more significant a risk, the more effort you should make to eliminate it through your design. Significant hazards are not necessarily those that pose the greatest risk but those that may not necessarily be obvious, are unusual, or are likely to be difficult to manage effectively.

To implement effective design risk management, you need to understand the difference between a hazard and a risk.

A hazard is an object, substance, situation or activity that has the potential to cause someone harm. For example, a floor without adequate slip resistance is a hazard because slips and trips can cause injuries. Similarly, loud noise from construction site plant and equipment is a hazard because it can cause temporary or permanent hearing loss; and breathing in asbestos dust is a hazard because it can cause cancer.

A risk is the likelihood – that is, the probability or frequency – that a hazard will actually cause someone harm, together with a measure of its effect. For example, a floor without adequate slip resistance is more likely to cause harm in a public space, and it is more likely to cause a serious injury if it is used on a staircase or ramp.

Confirm the slip resistance of a floor finish using independent test data before you specify it.

When you are designing your project, think about the changing circumstances in which people may be faced by potential hazards. For example, when designing accessible roofs, think about the hazards and likelihood of harm faced by personnel that need unplanned access to the roof in an emergency on a cold, wet, dark Friday afternoon, as well as those undertaking planned maintenance in more clement conditions. You should also consider the different risks hazards may pose throughout the building's lifecycle. The hazards faced by occupants during the occupation and maintenance phase of your project are likely to pose a risk for considerably longer than those during the construction phase, so being mindful of building safety during this period is as important as during the construction phase.

Some of the hazards that you ought to consider include:

- **Fire** poses a significant hazard, not just once your project is completed but also during the construction phase. Each year there are hundreds of fires on construction sites, putting the lives of site workers and the public at risk. The HSE's publication HSG168 provides guidance on how to reduce the risk of fires on construction sites.[5] The Fire Protection Association (FPA) also provides guidance in their publication 'Fire Prevention on Construction Sites', often referred to as the Joint Fire Code, relevant to the design stage of buildings.

  Your project fire strategy should consider fire safety during the construction phase of the project, in particular measures to reduce and control the risk of fire before the permanent fire safety installations are complete. This is particularly important for buildings that are more vulnerable to fire during the construction phase, for example timber framed buildings, and where fire spread from the construction site might endanger the lives of people in adjacent properties or buildings. Modern methods of construction that rely on component parts that are left temporarily unprotected or exposed once in-situ on site (for example, timber frame or composite buildings panels), may need extra precautions at certain vulnerable times of the build to minimise the risk of fire.

  Most construction-related fires are the result of flying sparks from hot works that can become trapped in cracks, gaps, holes and other small openings where they can smoulder and cause the outbreak of fire. Specifying systems and materials that negate the need for hot works can contribute significantly to avoiding fires on site. If hot works are necessary, the risk can be managed using temporary fire suppression systems, observing fire watches for an hour following the hot works, or having the appropriate fire extinguisher available close by (listed from most to least effective).

Agree the principles of the project fire safety strategy for the permanent works with your client at concept design stage, or as soon as you start work on the project if appointed at a later work stage. As a minimum, the fire safety strategy should address fire safety measures relating to means of warning and escape, external fire spread and access and facilities for the fire service, as well as recording all key design decisions relating to fire safety.[6] We look at some of the regulatory requirements and guidance relating to fire safety in Chapter 5.

- **Falls from height** are one of the biggest causes of fatalities and major injuries in our industry. Common cases include falls from ladders and through fragile surfaces. Take a sensible approach to managing this hazard. Equipment to enable safe working at height should be appropriate for the planned task to be undertaken and their associated risks, including temporary scaffolding with a working platform and fixed access ladders for more prolonged work, or the use of a mobile scaffold tower or MEWP, where fixed access is unnecessary. The HSE provides useful guidance regarding some of the dos and don'ts of planning for work at height.[7] Bear in mind there may be some low-risk situations where common sense tells you no particular precautions are necessary.

- **Musculoskeletal disorders** (MSDs) include injuries and conditions that can affect the back, joints and limbs and effect construction workers more than in any other industry. They are caused by manual handling, repetitive work and awkward postures. You should consider how your design can prevent or minimise the risk of workers developing MSDs, including avoiding or reducing the need for manual handling. The HSE provides useful guidance on musculoskeletal disorders at work.[8]

Take a sensible
approach to designing
for work at height.

- **Legionella bacteria** can lead to a potentially fatal type of pneumonia called Legionnaires' disease that can be contracted by inhaling airborne water droplets containing the bacteria. Legionella bacteria thrive in stagnant water but are dormant below 20°C and do not survive above 60°C, therefore water should not be stored between these temperatures. Public health systems should be designed to keep pipework as short and direct as possible with adequate insulation to pipes and tanks, using fittings and materials that prevent contamination and do not encourage the growth of Legionella.

- **Dust**, particularly silica dust, poses the biggest risk to construction workers' respiratory health after asbestos. Silica is a natural substance found in construction materials such as bricks, tiles, concrete and mortar. Cutting, drilling, grinding and polishing materials releases fine particles of silica dust that cause respiratory disease. You can minimise the health risks to workers by considering whether specifying materials containing silica could be avoided or moderated. Often specifying such

Develop your design to minimise the need for cutting or working silica construction materials on site.

materials is unavoidable, so develop your design to minimise the need for cutting or working the material on site.[9]

- **Slips and trips** are the most common cause of injury in the workplace and can lead to other accidents, including falls from height. Minimise the risk of slips and trips by specifying floor finishes that have a low slip potential.[10] Provide suitable information and advice to your client regarding the cleaning and maintenance regime required to maintain the slip resistance of the floor finish you have specified as dirt or contamination will impact its performance. Over-zealous cleaning and polishing of a floor will reduce its surface roughness and undermine its slip resistance with the potential to create a hazard where there was not one before. Confirm the slip resistance of a floor finish before you specify it, using independent test data to confirm the floor's slip potential in both the wet and dry states.[11] Be cautious of relying on applied finishes to achieve an adequate slip resistance for new floors, particularly finishes that will be applied on site, where inadequate quality control and cleaning or maintenance may undermine the required performance of the floor.

Consider how permanent works will be constructed and the impact this may have on the design, construction and dismantling of any temporary works.

- **Temporary works** used to provide provisional support or protection to the permanent works during the construction of your project require correct design, installation and maintenance to prevent and mitigate risks during construction. Whilst responsibility for the design and installation of temporary works may be delegated to a specialist and/or the contractor, you still need to understand and consider the requirement for, and practicality of, employing temporary works. Think about how the permanent works will be constructed and whether these may impact on how any temporary works can be safely constructed and dismantled without compromising your design. Consider whether it may be safer and/or more economical to sacrifice the temporary works and leave them in place following construction of the permanent works, which is often necessary for complex below-ground and sub-structure works.

Red, amber, green lists (commonly known as RAG lists) provide helpful guidance regarding the nature and type of hazards and risks to consider as part of your design risk management. A RAG list may also be used as a management or control tool by your client, or their principal designer or principal contractor, as part of the safety strategy and as a way of monitoring and communicating details regarding residual risks on your project. Industry guidance for designers produced by the Construction Industry Training Board (CITB) and HSE provides a good example of a RAG list.[12]

The red lists include hazardous procedures, products and processes that you should eliminate from your project wherever possible. For example, this could include proceeding with your design without adequate pre-construction information, designing glazing that cannot be accessed for cleaning or replaced safely, or designing structures that do not allow for fire safety and containment during construction.

Amber lists include products, processes and procedures to be eliminated or reduced, as far as possible, and only specified if unavoidable. Incorporating amber items in your design requires you to provide relevant information to the principal contractor and client regarding any residual risks. Examples of amber items include the specification of heavy building

> Regularly review your design and employ the general principles of prevention to reduce red and amber procedures, products and processes.

blocks (i.e. blocks in excess of 20kg), designing large and heavy glazing panels, and the specification of solvent-based coatings.

Green lists include products, processes and procedures that you should aim to include in your design. For example, provide edge protection where there is a foreseeable risk of falls from height, design practical and safe methods for window cleaning (such as from the inside), specify floor finishes to minimise the risk of slips and trips during use and maintenance, and design for off-site fabrication and prefabricated elements to minimise site hazards.

Carry out regular reviews of the red, amber and green aspects of your design to consider how to adapt or modify your design to reduce the red and amber elements and increase the green elements. The most effective way to achieve this is to employ the general principles of prevention.

4.2 General principles of prevention

You have a legal duty under the Management of Health and Safety at Work Regulations 1999 (Management Regulations) and the Construction (Design and Management) Regulations 2015 (CDM Regulations) to consider the general principles of prevention as part of your design risk management strategy. The principles of prevention, which apply to all industries, not just construction, provide a framework to identify and implement a proportionate design risk management strategy to control foreseeable risk.

Simply put, you employ the principles of prevention at all stages of your design development, from the beginning of the concept design stage to the completion of the technical design stage, to avoid putting people at risk.

The general principles of prevention, as specified in Schedule 1 of the Management of Health and Safety at Work Regulations 1999,[13] are:

1. **Avoid risks**: Consider the potential hazards created by your design and aim to avoid these by making variations to your design.

For example, replace areas of fragile roofing with load-bearing or impact-resistant construction to avoid falls from height, and specify pre-formed brick or blockwork to avoid the need for site cutting and exposure to dust.

2. **Evaluate the risks that cannot be avoided**: Consider the likelihood and severity of harm posed by the hazards that you cannot avoid, determining how much of a risk they pose to the people that will be constructing, maintaining, occupying and demolishing your project. Agree with your client what the tolerable level of risk is in the context of your project. Remember that what is acceptable may vary for different areas of the project, particularly if parts of your project will be accessible to members of the public.

3. **Combat the risks at source**: Adapt your design to avoid the need for further intervention in the future. For example, design glazing at height so that it can be safely cleaned from inside specifing reversible windows with restrictors and/or friction stays with appropriate guarding or sill heights to deal with the risk of falls from height. Design and specify the layout and size of fixed panes and spandrel panels in accordance with the maximum reach suitable for cleaning and maintenance from inside the building. This will minimise the risk of falls from height and avoid the need for specialist cleaning equipment. Bear in mind the maximum reach may need to include the window frame and sill as well as the glazed area.

4. **Adapt the work to the individual**: This is especially pertinent in the design of workplaces. Agree with your client the choice of work equipment and the choice of working and production methods, with a view to alleviating monotonous work and work at a predetermined work rate to reduce ill effects on the health of workers. Consider the design and specification of construction materials to avoid repetitive manual handling, for example by specifying a prefabricated, panelised wall system instead of large areas of masonry construction.

5. **Adapt to technical progress**: Whilst the basic process of construction has not changed significantly since the Middle Ages, the technical processes and methods used have. Advances in construction technology can improve quality, efficiency, value for money and sustainability as well as safety. Building Information Modelling (BIM)

modern methods of construction (MMC), modular construction, off-site manufacturing, prefabrication and preassembly, smart technology, robotics and GPS-controlled equipment are all examples of technology that can be utilised to improve safety on your project.

6. **Replace the dangerous by the non-dangerous or the less dangerous**: Consider the hierarchy of design solutions available to you. For example, when you are agreeing a maintenance strategy with your client, rope access may be discussed as an acceptable solution, but a less dangerous solution would be to design your building to be accessible at every storey by utilising roof terraces or balconies for safe level access and specifying windows that can be cleaned from the inside, or by incorporating a building management unit (BMU) into your design.

7. **Develop a coherent overall prevention policy**: Discuss and agree with your client a safety strategy for the project that considers the use of technology, organisation of work, working conditions, social relationships and the influence of factors relating to the working environment.

8. **Give collective protective measures priority over individual protective measures**: Any design solution that protects everyone from harm without the need for them to take any particular action is safer than one that only protects individuals or a small number of people and/or requires those people to undertake special training or a particular action or behaviour to ensure their safety. Accessible roofs with suitable permanent edge protection provide collective protection to everyone accessing the roof. Roof access that relies on a lanyard or fall prevention safety system for safety will only provide protection to a small number of individuals. A system-based solution will necessarily require access to be restricted to site personnel that have appropriate training to use the particular system specified, and for the system to be effective it relies on such site personnel actually making the effort to use it.

9. **Give appropriate instructions to employees**: Ensure that you provide clear, concise and relevant information to all those who may require it regarding any hazards that require a safety management plan.[14]

Advise your client where appropriate instructions or control measures may need to be identified or implemented by the building users.

If, during your design development, you find it is not reasonably practicable to design-out a risk despite applying the principles of prevention, provide adequate information regarding the residual risks to ensure that proportionate measures can be taken to control them at source. For example, advise your client where appropriate instructions or control measures may need to be identified or implemented by the building users and provide information about those risks to the principal designer.

Under The Regulatory Reform (Fire Safety) Order 2005 and The Higher-Risk Buildings (Management of Safety Risks etc) (England) Regulations 2023 the responsible person and/or the accountable person, as relevant, are legally obliged to adopt a hierarchy of principles to avoid building safety risks that are similar (but not identical) to the principles of prevention. To assist the responsible and/or accountable person to discharge their duties you should ensure that you share relevant information regarding your design risk management, in particular those hazards that you are aware they will need to manage in accordance with their statutory obligations.

4.3 Effective communication and coordination

You have a duty to coordinate your work with your client and members of the project, construction and management teams to ensure issues relating to risk or safety are effectively communicated to anyone affected by the building or building work. This may include not only your client, but also legal dutyholders, other project team members, occupants (including residents) and the emergency services.

As a minimum, you should contribute to the recording of, development, collection, organisation and sharing of information about your project's design, construction, operation, maintenance and refurbishment throughout the building lifecycle to ensure the golden thread of information is preserved. This will include obtaining, recording, updating and sharing information about your project and safeguarding and keeping the information secure. We consider your legal obligations under the

Building Safety Act 2022 to maintain the golden thread of information in more detail in Chapter 7.

Discuss and agree with your client their strategic requirements at the brief and concept stages of your project. Make sure you have documented and agreed project strategies for all building safety information in place from the outset of the project, for example, fire safety, structural safety, public health, cleaning and maintenance strategies. If you join a project after completion of the brief or concept design stage, take time to familiarise yourself with your client's brief and existing safety strategies, or prioritise putting these in place if they have not already been agreed.

You need to understand your client's attitude to risk and, relative to their knowledge and experience, explain to them (or remind them of) their duties to ensure they understand how important a client's role is and their potential to positively influence the safety outcomes of your project. If your client lacks experience or appears to have a relaxed attitude to health and safety, you need to take more time to advise them of their legal duties. This is particularly important for domestic clients for whom your project may be the first and only construction project they have commissioned. For every client you work with, including repeat clients, ensure that they understand that they need to meet certain standards and explain the potential implications of not doing so; not just the implications for them as a client (for example, prosecution and reputational damage) but also for the project and everyone involved in the project (for example, injury or death of personnel and delays and additional costs to the project).

Discussing and communicating health and life safety matters on your project should not be done separately to the design process. The most effective way to embed design risk management in the design process is to deal with it as part and parcel of client and design team meetings. This encourages regular and effective engagement with the whole design team. If you are appointed as the lead designer or lead design consultant, include regular discussion and information exchanges regarding relevant aspects of design risk management as part of your design coordination and client meetings, ensuring that the outcomes of such discussions are recorded to

Deal with design risk management as an integral part of your regular client and design team meetings.

capture relevant information regarding the decision-making process and any actions arising.

Deal with health and safety as you would any other aspect of your design development, on an ongoing basis. A design risk management review is not a task to be undertaken, ticked off and then forgotten. As your design develops, new hazards may present themselves or the level of risk posed by existing hazards may change. You need to address this on an ongoing basis and the most efficient and effective way to do this is to develop habitual behaviour amongst your design team to discuss and address hazards and risks as they arise, in a proportionate way and in conjunction with other competing project risks and opportunities.

Take time to produce clear and concise records of your design risk management discussions, including the agreed outcomes. Identify and agree with your client the records that should be kept and how those should be retained. This might be in the form of meeting minutes, annotations or mark-ups on drawings or digital models, or monthly reports, depending on the approach agreed with your client as part of your agreed health and safety strategy. Whatever is agreed, it is important that all members of the design and construction team take a consistent approach and that records are clear, concise, balanced, factual and proportionate to the risks on your project. This will ensure appropriate and relevant life safety information is identified, obtained and managed in a way that ensures it is accessible to anyone affected by the building and building work. This exercise may also require you to assist your client to identify and highlight missing information, which may involve consultation with existing occupants, including residents, if your project involves work to an existing building.

The Building Safety Act 2022 requires your client to use a digital strategy and system to coordinate and communicate design risk management information. To ensure that this is effective, you may need to assist your client to ensure that all members of the design, construction, occupation and maintenance teams (including any residential occupiers) have the competence and capability to engage effectively with such a digital strategy, including an appropriate level of access to, and knowledge of, the chosen software. A consistent approach to managing health and safety information is crucial to ensuring that all relevant information is available to those who need it and at the time they need it. Information should be coordinated at every stage of your project to ensure that it is all readily available in one place in a single, user-friendly format.

Employing the principles of prevention as part of your design development and approach to design risk management targets the elimination, so far as reasonably practicable, of all foreseeable risks associated with the lifecycle of your project. Despite this, there will always be risks that it is not possible, practicable or desirable to eliminate, considering your other desired project outcomes.

When you decide to reduce or control a risk, rather than eliminate it, inform your client and your client's professional team – including the contractor – regarding the residual risks that remain because of your decision and provide information about those risks to the principal designer. In doing so, the nature and level of information you provide should be project-specific,

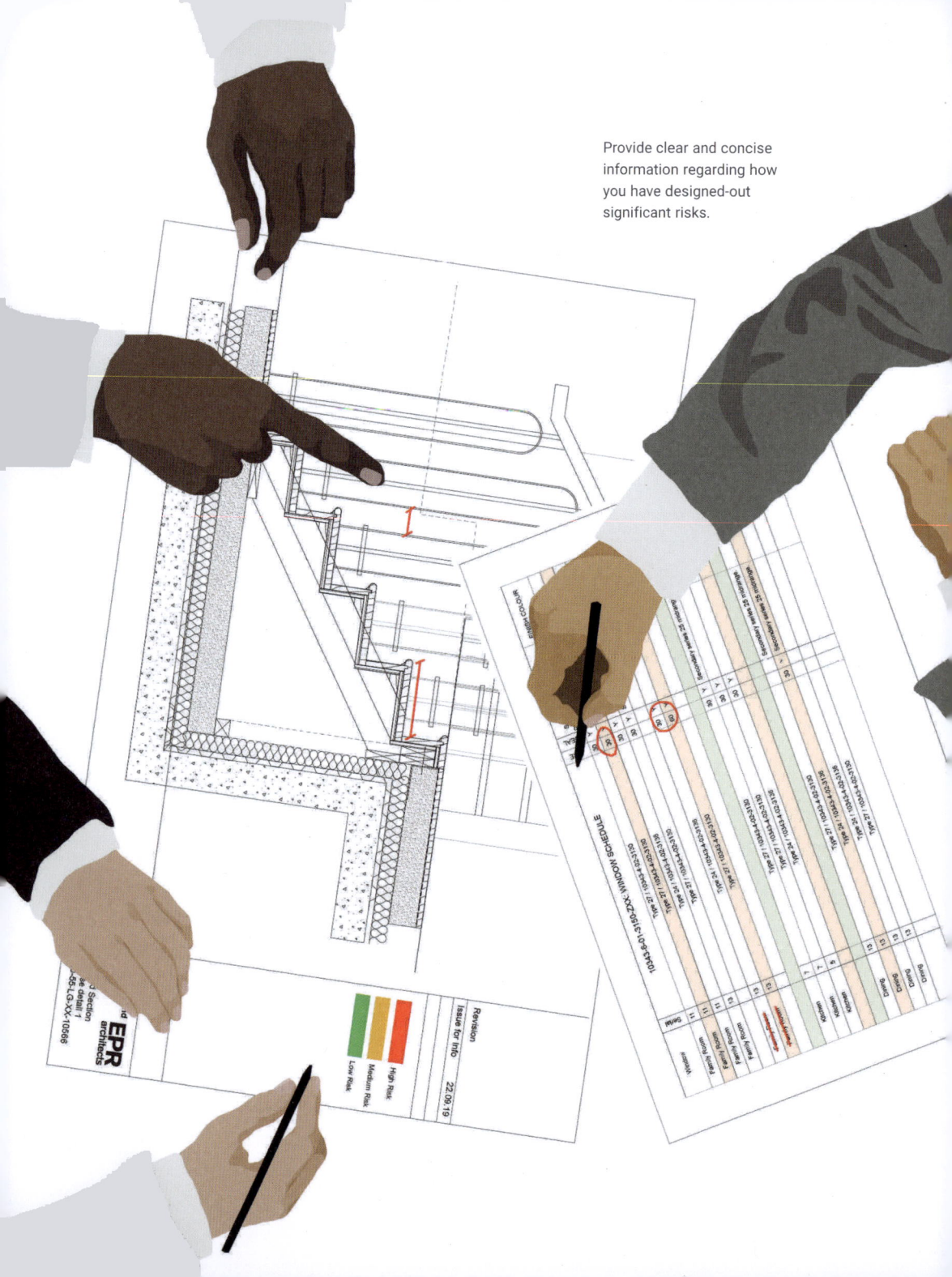

Provide clear and concise information regarding how you have designed-out significant risks.

clear, concise and proportionate to each risk. Do not use generic schedules or information copied from another of your projects. Whilst reviewing precedents may be helpful to decide what might be suitable in the context of your current project, they should not be taken to confirm the acceptability of your proposals. Do not include irrelevant information that will obscure the important information your client and/or their contractor requires to control the risks specific to your project.

The information you provide should focus on significant or unusual residual risk. Do not provide information regarding insignificant risks or those arising from routine construction or maintenance activities, of which a competent contractor or building manager ought to be aware. If you do, there is a danger the relevant information could be overlooked.

Carefully consider what information you should provide regarding risks that you have eliminated through your design risk management. Retaining details of your risk assessments for yourself will be useful to record your approach to the design risk management and decision-making process but may be of little value to your client and/or their contractor provided a hazard no longer poses a risk. Provide clear and concise information regarding how you have designed-out significant risks, particularly if responsibility for your design is passed to another designer and/or you are not retained to provide services in connection with your project through to completion. If it is foreseeable that a design variation by others later in the project could re-introduce a significant risk that you have previously eliminated, consider whether it may be appropriate to provide information regarding the positive design decisions you made to help the new design team to evaluate whether circumstances may have changed and whether it is still reasonably practical to eliminate the risk.

CHAPTER 5:
STATUTE, GUIDANCE, COMPETENCE AND CODES OF CONDUCT

To fulfil our architectural duties professionally we need to understand the current minimum legal requirements, in terms of professional duties and design, imposed by government legislation and codes of conduct. However, be mindful, these only provide a benchmark for minimum standards. Good practice and good-quality, sustainable, accessible and safe design may require you to exceed statutory minimums.

In this chapter, we consider the standards with which you need to be familiar:

5.1   Statute and the regulatory environment
5.2   Statutory and non-statutory guidance
5.3   Industry competence requirements and codes of conduct
5.4   Demonstrating competence

5.1 Statute and the regulatory environment

We work in a regulated industry with rules that mandate the minimum standards and duties regarding how we manage our work as it relates to the health, safety and well being of others. These statutory rules comprise primary and secondary legislation.

The extent to which legislation is applicable across the United Kingdom varies.[1] For the purposes of this guide, guidance provided is applicable to England, unless noted otherwise. You should confirm the extent to which legislation is applicable or differs in Wales, Northern Ireland and Scotland prior to undertaking projects in these jurisdictions.

• Primary legislation typically takes the form of an Act of Parliament and is the general term used to describe the main laws passed by the legislative bodies of the UK.

- Secondary legislation is law created by ministers (or other bodies) under powers given to them under the primary legislation. Secondary legislation often takes the form of regulations or statutory instruments.

The three pieces of primary legislation that you need to be familiar with are the **Health and Safety at Work etc. Act 1974** (HSWA),[2] the **Building Act 1984**[3] and the **Building Safety Act 2022** (BSA).[4]

The HSWA covers occupational health and safety in Great Britain. The Health and Safety Executive (HSE) is responsible for enforcing the HSWA, which sets out the general duties that employers have towards employees and members of the public, the duties employees have towards themselves and to each other, and the duties certain self-employed workers have towards themselves and others. There are several regulations that are enacted under the HSWA. Some of those that you need to be familiar with include:

- **The Confined Spaces Regulations 1997**, which defines what a confined space is and imposes obligations on employers to protect employees who may be working in confined spaces.
- **The Construction (Design and Management) Regulations 2015** (CDM Regulations), which impose duties on those responsible for the construction process, from concept to completion, to ensure projects are carried out in a way that secures health and safety.
- **The Control of Asbestos Regulations 2012**, which impose a duty to manage asbestos and asbestos-containing materials (ACMs) in non-domestic premises on every person with control over, or an obligation with respect to, maintenance or repair of that premises in order to protect anyone using or working in the building from health risks associated with exposure to asbestos.
- **The Management of Health and Safety at Work Regulations 1999** (Management Regulations), which make explicit requirements for employers to satisfy the requirements of the HSWA.
- **The Manual Handling Operations Regulations 1992** (MHOR), which place obligations on employers to manage the risks of manual handling faced by their employees.

If you are an employer, manage the risks of manual handling faced by your employees.

- **The Regulatory Reform (Fire Safety) Order 2005** (Fire Safety Order), which requires that a responsible person with control of a building takes reasonable steps to manage and maintain the building to reduce the risk from fire and makes sure people can escape safely from a building in the event of a fire.
- **The Work at Height Regulations 2005** (WAHR), which apply to all work at any height (even if it is at or below ground level) where there is a risk of a fall that may cause injury.
- **The Workplace (Health, Safety and Welfare) Regulations 1992**, which encompass a wide range of fundamental health, safety and welfare issues that are applicable to most workplaces, excluding construction sites.

The **Building Act 1984** is intended to secure the health, safety and welfare of people who may use buildings or might otherwise be affected by buildings or matters connected with buildings. Each national jurisdiction sets its own building regulations for building work. In England the **Building Regulations 2010**[5] are enacted under the Building Act and are a set of national building standards that apply to the design and construction of the majority of new buildings and alterations to existing buildings, including projects involving a material change of use.[6]

The primary purpose of the Building Regulations is to ensure that all building work is carried out to a standard that ensures the health and safety of building users, as well as standards for energy conservation, building access and security.

Regulation 4, 'Requirements relating to building work', of the Building Regulations makes it a legal requirement that building work is carried out so that it complies with the applicable functional requirements set out in Schedule 1 of the Building Regulations in England.

Schedule 1 is arranged in Parts A to S and each part includes a description of the legal minimum requirement or requirements that building work is obliged to meet. These are known as the 'functional requirements' and must be met either through the design or construction of the building work and are described in terms of function rather than form. This allows you the flexibility to determine the most appropriate design solution for your project that will meet the required function rather than being restricted to the use of a prescribed design solution.

- **Part A: Structure**
  - **A1 Loading**
  - **A2 Ground movement**
  - **A3 Disproportionate collapse**
- **Part B: Fire Safety**
  - **B1 Means of warning and escape**
  - **B2 Internal fire spread (linings)**
  - **B3 Internal fire spread (structure)**
  - **B4 External fire spread**
  - **B5 Access and facilities for the fire service**
- **Part C: Site preparation and resistance to contaminants and moisture**
  - **C1 Site preparation and resistance to contaminants**
  - **C2 Resistance to moisture**
- **Part D: Toxic substances**
  - **D1 Cavity insulation**
- **Part E: Resistance to the passage of sound**
  - **E1 Protection against sound from other parts of the building and adjoining buildings**

- – E2 Protection against sound within a dwelling house, etc
- – E3 Reverberation in common internal parts of buildings containing flats or rooms for residential purposes
- – E4 Acoustic conditions in schools
- **Part F: Ventilation**
  - – F1 Means of ventilation
- **Part G: Sanitation, hot water safety and water efficiency**
  - – G1 Cold water supply
  - – G2 Water efficiency
  - – G3 Hot water supply and systems
  - – G4 Sanitary conveniences and washing facilities
  - – G5 Bathrooms
  - – G6 Food preparation areas
- **Part H: Drainage and waste disposal**
  - – H1 Foul water drainage
  - – H2 Wastewater treatment systems and cesspools
  - – H3 Rainwater drainage
  - – H4 Building over sewers
  - – H5 Separate systems of drainage
  - – H6 Solid waste storage
- **Part J: Combustion appliances and fuel storage systems**
  - – J1 Air supply
  - – J2 Discharge of products of combustion
  - – J3 Warning of release of carbon monoxide
  - – J4 Protection of building
  - – J5 Provision of information
  - – J6 Protection of liquid fuel storage systems
  - – J7 Protection against pollution
- **Part K: Protection from falling, collision and impact**
  - – K1 Stairs, ladders and ramps
  - – K2 Protection from falling
  - – K3 Vehicle barriers and loading bays
  - – K4 Protection against impact with glazing

- – K5 Additional provisions for glazing in buildings other than dwellings
- – K6 Protection against impact from and trapping by doors
- **Part L: Conservation of fuel and power**
  - – L1 Conservation of fuel and power
  - – L2 On site generation of electricity
- **Part M: Access to and use of buildings**
  - – M1 Access and use of buildings other than dwellings
  - – M2 Access to extensions to buildings other than dwellings
  - – M3 Sanitary conveniences in extensions in buildings other than dwellings
  - – M4(1) Category 1: Visitable dwellings
  - – M4(2) Category 2: Accessible and adaptable dwellings
  - – M4(3) Category 3: Wheelchair user dwellings
- **Part O: Overheating**
  - – O1 Overheating mitigation
- **Part P: Electrical safety**
  - – P1 Design and installation of electrical installations
- **Part Q: Security – Dwellings**
  - – Q1 Unauthorised access
- **Part R: Physical infrastructure for high-speed electronic communications networks**
  - – R1 In-building physical infrastructure
- **Part S: Infrastructure for the charging of electric vehicles**
  - – S1 The erection of new residential buildings
  - – S2 Dwellings resulting from a material change of use
  - – S3 Residential buildings undergoing major renovation
  - – S4 Erection of new buildings which are not residential buildings or mixed-use buildings
  - – S5 Buildings undergoing major renovation work which are not residential buildings or mixed-use buildings
  - – S6 The erection of new mixed-use buildings and mixed-use buildings undergoing major renovation

Details of the functional requirement for each part can also be found in the green text boxes at the start of each relevant section of the Approved Documents guidance that accompany the Building Regulations.

Regulation 7, 'Materials and workmanship', of the Building Regulations makes it a legal requirement that building work shall be carried out with adequate and proper materials that are appropriate for the circumstances in which they are used, are adequately mixed or prepared and are applied, used or fixed so as to adequately perform the functions for which they are designed, and that building work shall be carried out in a professional manner.

Regulation 38, 'Fire safety information', of the Building Regulations applies with respect to building work where Part B of Schedule 1 imposes a requirement in relation to work on a relevant building. The regulation makes it a legal requirement that the person carrying out the work shall give fire safety information to the responsible person not later than the date of completion of the work, or the date of occupation of the building or extension, whichever is the earlier.

A 'relevant building' in relation to Regulation 38, is a building to which the **Regulatory Reform (Fire Safety) Order 2005** applies or will apply after completion of building work. This generally means any building that is or may be a workplace, excluding domestic premises.

'Fire safety information' means information relating to the design and construction of the building or extension, and the services, fittings and equipment provided in, or in connection with, the building or extension, which will assist the responsible person to operate and maintain the building or extension with reasonable safety.

The 'responsible person' means, in relation to a workplace, the employer, if the workplace is to any extent under the employer's control. In relation to any other premises, it is the person who has control of the premises (as occupier or otherwise) in connection with the carrying on by that person of a trade, business or other undertaking (for profit or not), or; the owner, where the person in control of the premises does not have control in connection with the carrying on by the person of a trade, business or other undertaking.[7]

The **Building Safety Act 2022** (BSA)[8] was introduced by the UK Government in response to Dame Judith Hackitt's Independent Review of Building Regulation and Fire Safety.[9] The duties under the BSA apply to all buildings subject to building control approval in England, including extensions, refurbishments and buildings subject to a change of use. The BSA includes enhanced statutory duties that are applicable to higher-risk buildings.

As well as the specific duties for designers under the BSA itself there are a number of regulations enacted under the BSA that you need to be familiar with. These include:

- **The Higher-Risk Buildings (Descriptions and Supplementary Provisions) Regulations 2023,**[10] which provides details regarding the definition of higher-risk buildings under the BSA, including details of exclusions from the definition and how to determine the height of a building within the scope of the regulations.
- **The Building (Approved Inspectors etc. and Review of Decisions) (England) Regulations 2023,** [11] which amends the Building (Approved Inspectors etc.) Regulations 2010 to support the higher-risk building control regime.
- **The Building Regulations etc. (Amendment) (England) Regulations 2023,**[12] which amends the Building Regulations 2010 to strengthen the regulatory regime for all building design and construction and to support the higher-risk building control regime including Part 2A 'Dutyholders and competence' requirements that apply to all building work, including that undertaken on higher-risk buildings.
- **The Building (Higher-Risk Buildings Procedures) (England) Regulations 2023,**[13] which provides the detail of the building control regime for higher-risk buildings including specifying the procedural building regulation requirements when a new higher-risk building is being designed and constructed or when building work is being done to an existing higher-risk building.

The dutyholders and competence requirements imposed on designers by Part 2A of the Building Regulations are applicable to the 'relevant requirements', which means 'to the extent relevant to the building work or design work in question, the requirements of regulations 4, 6, 7, 8, 22, 23, 25B, 26, 26A, 28, 36, 41(2)(a), 42(2)(a), 43(2)(a), 44A, 44ZA, 44ZC and 44D to 44I and Schedule 1 of the Buildings Regulations 2010'. This means that if you are working on a project where the functional requirements of the Building Regulations are applicable and your design is subject to building control approval, you will also be subject to, and need to be familiar with, the designer's duties and competence requirements under the building regulations.

5.2 Statutory and non-statutory guidance

To assist you to discharge your statutory duties appropriately various government departments, agencies and industry bodies produce memoranda, Approved Documents, Approved Codes of Practice, standards and general guidance that provide advice to help you to understand your legal duties and how you might discharge those duties, either through the design of your project or your actions as a designer, principal designer and an employee and/or employer.

Industry guidance may be statutory or non-statutory. Statutory guidance has a legal status granted under applicable legislation, for example the Approved Documents under the Building Regulations. Non-statutory guidance generally does not have a legal status, however, some non-statutory guidance may be produced or approved by the relevant Secretary of State, which confers a special legal status on it, for example Approved Codes of Practice (ACoPs), published by the Health and Safety Executive (HSE). If you follow statutory guidance, or an ACoP, you will normally be doing enough to comply with the law. However, you are not obliged to follow statutory guidance and you may use alternative methods to meet your statutory obligations based on non-statutory guidance.

Examples of statutory and non-statutory documents that provide guidance regarding how you might discharge your duties under the primary and secondary legislation include:

Explanatory notes and memoranda, which are published by the government department responsible for producing legislation. Explanatory notes generally accompany primary legislation – Acts of Parliament, and set out what an act sets out to achieve. Explanatory memoranda accompany secondary legislation – statutory instruments, and set out a brief statement of the purpose of each statutory instrument while providing information about its policy objectives. Both aim to make acts and statutory instruments accessible to readers who are not legally qualified.[14]

Approved Documents, which are published by the government and provide statutory guidance on ways to meet the functional requirements of the Building Regulations.[15] The documents contain general guidance on the performance expected of materials and building work to comply with the Building Regulations as well as practical examples and solutions on how to achieve compliance for some of the more common building situations. There is an Approved Document for each of the parts of Schedule 1 of the Building Regulations, as well as an Approved Document providing guidance regarding material and workmanship to comply with Regulation 7. The guidance in the Approved Documents is informative only, but is periodically reviewed and updated, particularly when changes to the Building Regulations and/or Schedule 1 are proposed or introduced. You should ensure that you are always referring to the edition of each Approved Document that is relevant to the design of your project. You may employ an alternative means of satisfying the minimum functional requirements of the Building Regulations, for example, by designing your building in accordance with the guidance in relevant British or European Standards.

Government departments and agencies also publish non-statutory guidance from time to time to supplement or clarify statutory guidance,

Approved Documents A to S provide statutory guidance on ways to meet the functional requirements of the Building Regulations.

128

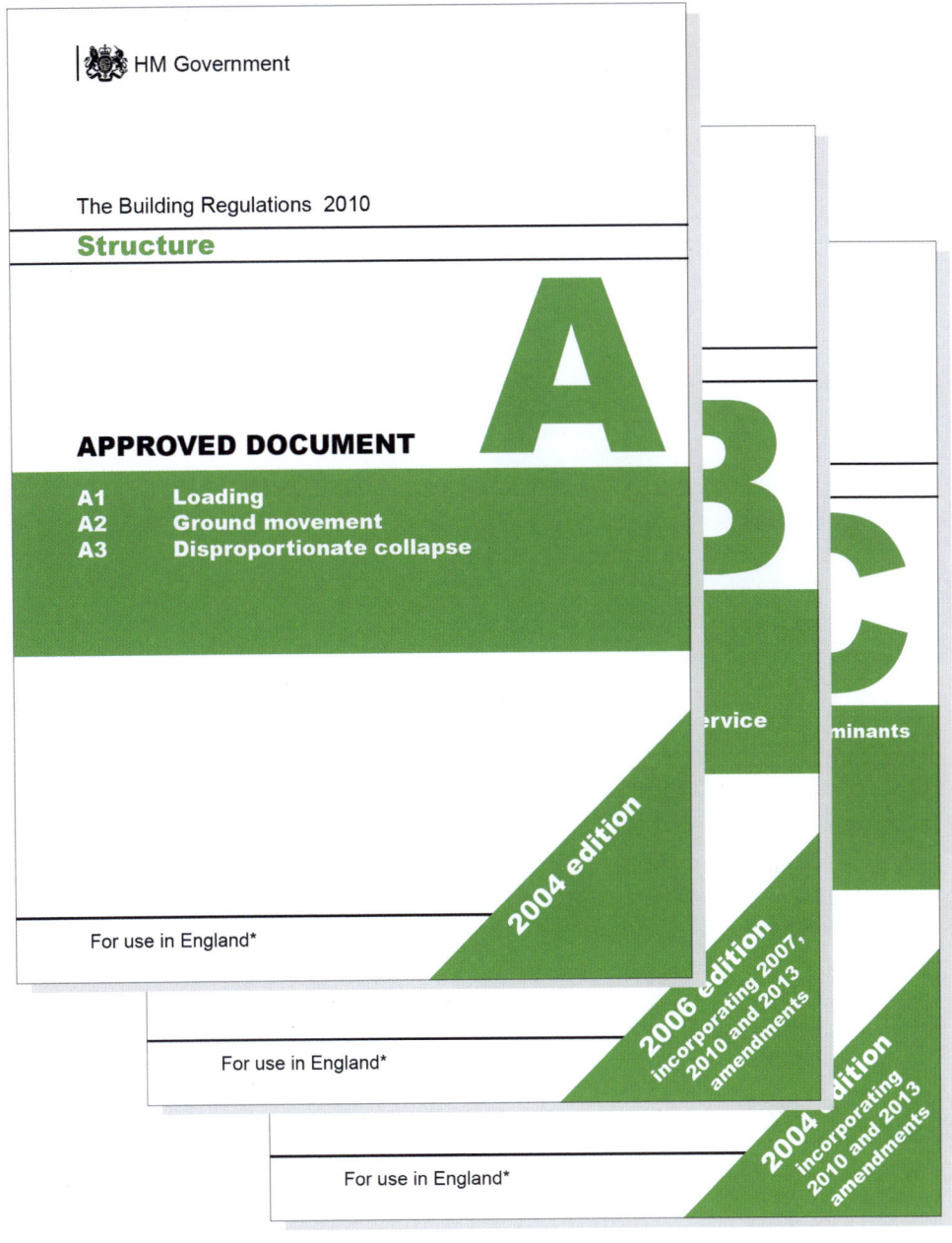

which is intended to assist you to comply with legislation. Examples of this include the 'Building (Amendment) Regulations 2018: frequently asked questions', which provides clarifications regarding the application of the amendments to the building regulations that banned the use of combustible materials in the external walls of buildings,[16]and the HSE's Building Control overview of Gateways 2 and 3 for the higher-risk building control regime [17].

British and European Standards contain non-statutory guidance produced by, for and on behalf of industry and are published by the British Standards Institute (BSI).[18] Following the guidance in these standards is one way to demonstrate that you have discharged your statutory duties. For example, guidance in relevant standards may be used as an alternative to the Approved Documents to demonstrate compliance with the functional requirements of the Building Regulations. Bear in mind that British Standards take the form of guidance and recommendations, which you need to take the time to understand to ensure that you use the guidance correctly into your design and specifications and are able to justify any proposal that deviates from the recommendations. Do not rely on British Standards as if they are a specification in themselves.

British Standards that are pertinent to health, safety and well being that you should be familiar with include:

- **BS 7974 Application of fire safety engineering principles to the design of buildings – Code of practice**. This standard sets out a framework for fire safety engineering principles in building design. The guidance provides recommendations and guidance for life safety and property and environmental protection with respect to fire safety and covers both new and existing buildings.
- **BS 8560 Code of practice for the design of buildings incorporating safe work at height**. This standard sets out the considerations and issues of incorporating systems for safe work at height into the design of buildings. The guidance recommends early involvement by the design team and is aimed at designers and those involved in construction, inspection, cleaning and maintenance.

British Standards provide non-statutory guidance on ways to meet the functional requirements of the Building Regulations.

BS 9991:2015
Incorporating Corrigendum No.1

BSI Standards Publication

Fire safety in the design, management and use of residential buildings - Code of practice

bsi.

- **BS 9991 Fire safety in the design, management and use of residential buildings – Code of practice**. This standard provides recommendations and guidance for the design, management and use of residential buildings, focusing on ensuring the fire safety of all people. The standard applies to dwellings, residential accommodation blocks and specialised housing. Guidance covers designing means of escape, active fire protection, construction design, stairs and exits, HVAC, ancillary accommodation to flats and maisonettes, access and fire-fighting facilities, and building works.
- **BS 9999 Fire safety in the design, management and use of buildings – Code of practice**. This standard provides recommendations and guidance for the design, management and use of buildings, focusing on ensuring the fire safety of all people in and around buildings. The standard applies to new buildings and to alterations and extensions to buildings, as well as changes of use of an existing building. The guidance covers the entire life cycle of a building, using a risk assessment approach and risk profiles, and looks at means of escape and evacuation strategy, access and fire-fighting facilities, building structure (including load and non-load bearing elements) and special risk protection.
- **BS EN 13501-1 Fire classification of construction products and building elements**. This standard provides a procedure for classifying all construction products[19] according to their reaction to fire considered in relation to their end use application.[20] Products can be classified, according to their fire performance, into seven ratings: A1, A2, B, C, D, E and F, where A1 rated products are non-combustible and exhibit no sustained flaming when tested and F rated products are the most combustible, igniting within 20 seconds. A2, B, C and D rated products may also be classified according to their propensity to produce smoke (s1, s2 or s3) and/or flaming droplets/particles (d0, d1 or d2) once ignited, where an s3 or d2 rating indicates the worst performance with unlimited production of smoke or flaming droplets/particles.

Exercise caution when following the guidance in British and/or European standards to ensure that it remains relevant in the context of current legislation. Whilst the BSI regularly reviews and updates the standards that it publishes, the process to review and update a standard can be a lengthy one, particularly if it requires public consultation regarding the technical content. This can mean that the practical guidance in some standards may be out of step with current legislation. Always check that you are using the appropriate version of a standard to inform your design and that it remains applicable to the relevant statute.

The HSE publishes a wide range of non-statutory guidance documents relevant to many areas of health, safety and well being, many of which are pertinent to the construction industry and our role as designers of the built environment.

ACoPs published by the HSE, with the consent of the Secretary of State, provide practical advice and describe preferred or recommended methods that can be used (or standards to be met) to comply with regulations and the duties imposed by the HSWA. ACoPs have a special legal status. If you are prosecuted for breach of health and safety law, and it is proved that you did not follow the relevant provisions of an ACoP, you will need to show that you have complied with them in some other way, or a court may find you at fault.[21]

Examples of HSE guidance that you should be familiar with include:

- **HSG168**: Fire safety in construction, which explains how everyone involved in a construction project can comply with their legal duties relating to fire risks.
- **L143**: Managing and working with asbestos, which provides guidance for employers about work which disturbs, or is likely to disturb, asbestos.
- **L153**: Managing health and safety in construction, which provides guidance on the legal requirements for the CDM Regulations, including the responsibilities of the duty holders under the regulations.

As well as regularly consulting the guidance published by the government, the BSI and HSE, you should also be familiar with any relevant practical and technical guidance produced by industry for designers and specifiers relating to the safe design, construction and use of specific materials, products and/or systems.

Industry experts, organisations, manufacturers and suppliers produce a wealth of guidance that can be helpful to assisting you ensure that your designs and specifications comply with the functional requirements of the Building Regulations. Be mindful that you should exercise caution and request verification that any technical claims made, particularly by materials and products manufacturers and suppliers, are based on sound research and testing.

An example of industry guidance that you may find useful is the Centre for Window and Cladding Technology (CWCT) and Society of Façade Engineering (SFE) Fire Guidance, which proposes practical and pragmatic interpretations for some of the potentially conflicting interpretations of the amendments to Building Regulation 7 and Requirement B4.[22]

As part of your safety and design risk management strategy, agree with your client and other members of the design team at the outset of your project which forms of statutory and/or non-statutory guidance you will follow to demonstrate compliance with any legislative requirements.

You may find it necessary to use one piece of guidance to supplement another. For example, you might need to use BS 9991 to supplement the guidance in Approved Document B. However, do not pick and mix guidance from different sources, as it is essential that you adopt an integrated approach to your design.

Whilst non-statutory guidance is helpful in assisting you to understand your statutory duties, it is no substitute for taking the time to read and understand the statute itself. Likewise, you may find it helpful to seek advice from professional or legal advisers to help you navigate the various pieces of legislation. However, this always comes at the risk of advisers imposing an emphasis or bias on how to interpret or meet the intent of the legislation. To make sure you have a proper understanding of your statutory duties, always make time to read and digest the actual text of any legislation that is relevant to you, your clients and your projects.

5.3 Industry competence requirements and codes of conduct

As a designer, you have a general duty under the **Construction (Design and Management) Regulations 2015** (CDM Regulations) to ensure that you have the skills, knowledge and experience necessary to fulfil the role you are appointed to undertake in a manner that secures the health and safety of anyone affected by your project.

You also have a general duty as a designer under Part 2A 'Dutyholders and competence' of the **Building Regulations 2010** (as amended by the **Building Regulations etc. (Amendment) (England) Regulations 2023**) to ensure that you, and anyone working under your supervision or employment as an individual, are a competent designer, i.e. has the skills, knowledge, experience and necessary behaviours to design work so that the building work to which the design relates, if built, would be in accordance with all the relevant requirements of the Building Regulations. These duties do not apply if you are in training, provided that you ensure that you, and the design work you undertake, are adequately supervised by a competent person.

> For the purposes of Part 2A of the Building Regulations, 'necessary behaviours' include:
> (a) compliance with the relevant requirements including refusing to carry out any design work if the building work to which the design relates cannot be carried out in compliance with all relevant requirements;
> (b) cooperating with others in relation to your design work;
> (c) refusing to carry out design work which is beyond your skills, knowledge or experience and asking for the assistance of others where necessary.

The general requirement under the CDM Regulations and Part 2A of the Building Regulations to ensure competence also applies at practice level, so if you are running a practice, you must ensure that your practice has the organisational capability to ensure design work undertaken by your practice is in accordance with all the relevant requirements of the Building Regulations.

> For the purposes of Part 2A of the Building Regulations, 'organisational capability' means having appropriate management policies, procedures, systems and resources to ensure individuals working for your practice who are carrying out any design work comply with the competence requirements of Part 2A of the Building Regulations, either as designer or principal designers, and any individuals working for your practice who are in training to develop the necessary skills, knowledge, experience and behaviours are appropriately supervised.

Your legal duty to ensure you are competent applies to any design work on all buildings in England – whether or not the design involves a higher-risk building – and includes design work carried out on existing and occupied buildings.

Your level of competence should be proportionate to your role and responsibilities. For example, if you are a project architect leading the team you will require a greater level of competence than if you are an architectural assistant, or in training, working under supervision as part of the design team.

If you find yourself in a situation where you are no longer competent to carry out the design work you have been appointed to provide, you have a legal duty under Part 2A of the Building Regulations to notify the person who has asked you to carry out the design work, and your client's principal designer.

The Architects Act 1997 (as amended by the **Building Safety Act 2022**) includes provisions for the Architects Registration Board (ARB), acting in its regulatory capacity, to monitor the competence of architects throughout their registration, including the power to determine which practical experience or training should be assessed and how the assessment should take place to determine competence.[23]

The **ARB Code of Conduct** (ARB Code) sets out the standards of professional conduct and practice expected of all architects registered by the ARB. If you are a registered architect, you are expected to be guided in your professional conduct and professional work by the spirit of the ARB Code as well as by its express terms.[24]

Standard 2: Competence of the ARB Code states that you are expected to be competent to carry out the professional work you undertake, and to keep your knowledge and skills relevant to your professional work up to date. Complying with Standard 2 requires you to keep the knowledge and skills relevant to your professional work up to date, particularly in response to advancements in technology and methods within the built environment, or changes in regulatory guidelines.[25]

If the ARB determines that you do not meet the requirements set out by the ARB, or you have not met the requirements to a sufficient standard, the ARB may remove you from the register. The ARB may, at its discretion, extend the time you have to demonstrate your competence, particularly if you can demonstrate that you are undergoing training to satisfy the standard of competence required.

The **Royal Institute of British Architects** (RIBA), in its capacity as a professional membership body, publishes a **Code of Professional Conduct** (the RIBA Code), which sets out and explains the standards of professional conduct and practice which the RIBA requires of all its members. The RIBA Code applies to all members, whether they are working in traditional architectural practice or have followed a different career path, such as in a multidisciplinary organisation, academia or a construction company.[26]

The purpose of the RIBA Code is to promote good conduct and best practice. Members must at all times comply with all relevant legal obligations, but the RIBA Code does not seek to duplicate legal obligations.

Principal 2: Competence of the RIBA Code states that:

Members should continuously strive to improve their professional knowledge and skill. Members should persistently seek to raise the standards of architectural education, life-long learning, research, training and practice for the benefit of the public interest, those commissioning services, the profession and themselves. Members should strive to protect and enhance heritage and the natural environment.

If you are a member of the RIBA, your duties under the RIBA Code in connection with health and safety require you to:

- Be familiar and up to date with relevant codes of practice and guidelines that may be issued or endorsed by the RIBA from time to time, especially those concerned with health and safety, ethical practice, sustainability and protection of the environment (Principle 1.15).
- Have reasonable knowledge of, and abide by, all laws and regulations relating to health and safety as they apply to the design, construction and use of your building projects (Principle 6.1).
- Take reasonable steps to ensure that your clients, and those with whom you have a direct professional relationship, are aware of and understand their responsibilities under the laws and regulations described in 6.1 above (Principle 6.2).[27]
- Take reasonable steps to protect the health and safety of those under your direct control of instruction (Principle 6.3).

- Take reasonable steps to protect the health and safety of those carrying out, or likely to be directly affected by, construction work for which you are providing professional services. This includes clients and members of the public (Principle 6.4).
- Not enter into any contract which compromises your duty to protect health and safety (Principle 6.5).
- Notify your client if you become aware of anything that compromises, or may compromise, your duty to protect health and safety (Principle 6.6).
- Take appropriate action if you become aware of a decision taken by your employer or client which violates any law or regulation and that will, in your professional judgement, materially and adversely affect health and safety. This includes advising your employer or client against the decision, and/or refusing to consent to the decision, and/or reporting the decision to the local building inspector or other public official charged with the enforcement of the applicable laws and regulations, unless you are able to satisfactorily resolve the matter by other means (Principle 6.7).

To support its members to fulfil their commitment to life-long learning the RIBA has published 'The Way Ahead' an education and professional development framework. The framework signifies a new direction for architectural education and continuing professional development, with greater emphasis on health and life safety, the climate emergency and professional ethics.[28]

'The Way Ahead' includes a set of mandatory competencies, including health and life safety. Complying with these requirements is one way you will be able to demonstrate your competence to act as a designer under the statutory requirements of the **Building Safety Act 2022**.

5.4 Demonstating competence

If you are appointed to carry out design work in connection with a higher-risk building, whether that's a new building, work on an existing one or work that involves a change of use that creates a higher-risk building your client has a duty under Part 2A of the Building Regulations to take all reasonable steps to satisfy themselves that you fulfil your designer's competence duties. This obliges your client to keep a written record of the steps they took prior to your appointment to ascertain that you are competent and to submit a copy of this record to the Building Safety Regulator for approval as part of the building control approval application for your project.

To enable you to adequately demonstrate to your client that you have the skills, knowledge, experience and behaviours required of a designer under Part 2A of the Building Regulations, you may need to use a range of documents to enable you to adequately evidence relevant details of the different aspects of your competence.

An up-to-date CV is an obvious place to start. This should include, as a minimum, details of your academic and vocational qualifications, including details of what, where and when you obtained these, as well as details of any structured training that you have undertaken that is pertinent to your project or role, again including details of what, where and when you undertook this. Have copies of any certificates or other evidence of your qualifications and training available so that you can provide these to your client in support of your CV, if required. You should also include details of any professional registration or memberships that you have that are relevant to your design responsibilities.

Your CV also ought to include examples of your project experience that are relevant to the project you will be working on. This should include not only details of the project itself – for example, what it was, when it was constructed and how much it cost – but also, and more importantly, what role and responsibilities you had in respect of the design and delivery of the project. You may need to use a combination of projects to enable you to demonstrate the range and breadth of your competence but be careful to ensure that you keep this relevant to your client's current requirements. For example, if you are being appointed to provide designer services in connection with a domestic project do not use a CV that you have previously produced containing commercial projects for a developer and vice versa.

If you are being appointed to provide design work in connection with a higher-risk building, your CV should also include a statutory statement confirming that you have not been subject to any serious sanctions in respect of any safety breaches over the last five years prior to the date of your appointment in relation to any projects you have worked on. Unless of course you have been subject to such sanctions, in which case you must include details of these, including the circumstances that led to the action being taken against you and what you have done to address any shortcomings to ensure the same does not happen again.

Serious sanction means:
(a) the issue of a compliance notice which referred to contravention or likely contravention of a requirement of Part A (structure) or Part B (fire safety) of Schedule 1 of the Building Regulations 2010;
(b) the issue of a stop notice;
(c) conviction for any offence under –
 i. the Building Act 1984,
 ii. the Health and Safety at Work etc. Act 1974,
 iii. the Building Safety Act 2022,
 iv. the Regulatory Reform (Fire Safety) Order 2005;
(d) a report published by an inquiry under the Inquiries Act 2005 that finds your action or inaction resulted in one or more deaths or was likely to have been a contravention of any requirement of –
 i. the Building Act 1984,
 ii. Part A (structure) or Part B (fire safety) of Schedule 1 of the Building Regulations 2010,

iii. the Health and Safety at Work etc. Act 1974,

iv. the Building Safety Act 2022,

v. the Regulatory Reform (Fire Safety) Order 2005.

The project experience you include in your CV and your statutory statement will go some way to demonstrating you have the behaviours necessary to undertake your designer role competently. It would also be beneficial to have some form of personal statement that provides examples of situations or issues that you have addressed, which demonstrate that you are able to cooperate effectively with other designers, you understand the extent and limitations of your competence and have the confidence and willingness to decline work where this falls outside the scope of your competence and/or to seek the advice and support of others to address this. You should also include relevant examples of project experiences you may have had as a designer where it has been necessary for you to challenge the approach your client or their design team may have taken where you have realised that such design does not comply with the relevant requirements of the Building Regulations or may otherwise had detrimental safety outcomes for the project if left unchallenged.

You need to ensure that your competence to provide designer services remains current and up-to-date, and to be able to demonstrate the same. To this end, having a structured training plan and training record that is regularly reviewed and updated will be invaluable. You should use your training plan to regularly review and evaluate your competence, to identify gaps and to set objectives for improvement. This will enable you to plan relevant training, either structured or unstructured, to ensure that you address these training requirements. Include the objectives and outcomes you intend to achieve. Your training record will then provide you with a method of evaluating and recording the training you have undertaken to confirm that this has met the objectives and outcomes you required, or to identify whether further training may be necessary. Ensure that your plan and record include details of how regularly and when you last reviewed and updated them. This will enable you to use the documents to demonstrate to your client that you take evaluating, planning and maintaining your competence seriously and have a robust process in place to manage and assess this.

CHAPTER 6:
CONSTRUCTION (DESIGN AND MANAGEMENT) REGULATIONS 2015

As soon as you carry out design in connection with any construction project based in Great Britain, including projects for a domestic client, you assume legal duties under the **Construction (Design and Management) Regulations 2015** (CDM Regulations).[1] The CDM Regulations apply to all construction projects as a whole – from concept to completion – and there is no minimum or maximum threshold in terms of size, type or value at which the CDM Regulations apply.

In this chapter we consider the legal duties imposed on designers and principal designers under the CDM Regulations and consider how these compare to the duties under the Building Safety Act:

6.1   Regulation 8 – General duties
6.2   Regulation 9 – Designer duties
6.3   Regulation 11 – Principal designer duties
6.4   CDM Regulations and the Building Safety Act

Clients, principal contractors, contractors and workers also have duties under the CDM Regulations, but consideration of these, other than how they relate directly to designers and principal designers, is beyond the scope of this guidance.[2]

6.1 Regulation 8 – General Duties

You have general duties under Regulation 8, which apply to everyone working on a construction project.

The CDM Regulations impose different obligations on us as individual designers and on our practices as organisations. If you are employed in practice, a sole trader or self-employed, the obligations on a designer are the most relevant. If you are the owner, partner or director of a practice, both the obligations of a designer and of an organisation are relevant. Either way, it is important to understand the general duties of both and to ensure they are being discharged appropriately on your project.

In Chapter 5 we considered how you need to be competent to undertake your statutory duties. As a designer under the CDM Regulations this means you must have the appropriate skills, knowledge and experience to address the anticipated risks on your project and to complete the services you have been appointed to provide. You will need to determine for yourself, and demonstrate to your client, your relevant skills, knowledge and experience based on the complexity of each project on which you work. This should be done in a way that is specific and proportionate to the nature of each project, avoiding excessive, generic or duplicated paperwork. One way to demonstrate this is to produce a CV that that is tailored to your client's current project and provides relevant details of your qualifications, professional memberships and experience of working on projects of a similar nature, size and complexity.

As a practice, you must have the organisational capability to perform the role and undertake the tasks for which your practice has been appointed. You must be able to demonstrate to your client that you have policies and systems in place to set acceptable health and safety standards within your practice, as well as the resources and people to deliver these at project level. This includes being able to demonstrate that your employees and colleagues are aware of your procedures and that these procedures are being implemented effectively.

As a minimum, you need to be able to demonstrate knowledge of the following:

- How does your practice manage building safety training? How does it manage continuing professional development to ensure it has the necessary organisational capabilities, and your designers have the necessary skills, knowledge and experience? For example, by having training plans and records for all your designers and a training matrix for your staff to identify and record the relevant knowledge, skills and experience across your practice.
- What is your system for design risk management and what is the review process to confirm that the system is being implemented effectively at project level? For example, by having project control files and design risk management reviews for all of your projects and ensuring that these are regularly reviewed and updated at project and practice level to identify deficiencies and opportunities for improvement.

- What is your review process for checking that building safety is being considered as part of your design development? For example, by having regular reviews of your projects, ideally with someone outside of the project team, to verify that safety matters are being identified and addressed in the design.
- How do you manage and ensure effective internal and external design team communication and cooperation? For example, by having project tracker documents to record, monitor and signpost relevant project correspondence, meeting minutes and site inspection records.

Unless or until you have the necessary skills, knowledge and experience, or your practice has the necessary organisational capability, you must not accept or proceed with an appointment on a project.

Other general duties under Regulation 8 include the duty to:

- cooperate with anyone working in connection with your project; this includes cooperating with owners of adjoining properties as well as teams who may be working on adjoining developments
- provide clear, concise information or instructions in simple, comprehensible English and/or other languages where appropriate; this should be limited to providing information or instructions relating to your design and only to those who you are employed to advise
- report dangerous conditions; typically this would be limited to promptly alerting the person in control of the site of any unsafe working practices or conditions that you may have witnessed during a site inspection. You should avoid intervening directly on site unless someone is in immediate danger, and it is safe for you to do so.

6.2. Regulation 9 – Designer duties

In addition to the general duties under Regulation 8, Regulation 9 sets out the duties that relate to all designers working on construction projects in Great Britain, including designers working overseas. These duties apply to all projects and at all stages, including concept designs, feasibility studies and competitions or speculative work. Your designer duties apply irrespective of whether a principal designer has been appointed or not and regardless of your client's requirements.

You are deemed to be a designer under the CDM Regulations if you prepare or modify a design or arrange or instruct someone under your control to do so.[3] These duties apply regardless of the status or nature of your appointment or any agreement you have with your client, including design work you may carry out for no fee or for a friend or family member.

Design work includes drawings, design details, specifications and bills of quantities (including specification of articles or substances) relating to a structure, and includes calculations prepared for the purpose of a design.

As well as being aware of your duties as a designer, before you commence any design work on your project you have a duty to satisfy yourself that your client is aware of their duties under the CDM Regulations. To do this you need to be sufficiently aware of the client's duties, which are set out in Regulations 4 to 6 and include:

- Duties in relation to managing the project, including making sure enough time and sufficient resources are available for the duration of the project, to ensure that the construction work can be carried out, so far as is reasonably practicable, without risks to the health and safety of any person affected by the project; and that welfare facilities are provided for those carrying out construction work (Regulation 4).
- A duty to appoint a principal designer and a principal contractor, where it is reasonably foreseeable that more than one contractor will be working on the project at any time. If your client fails to make either or both appointments, the client must fulfil the role or roles (Regulation 5).
- Notifying the HSE about the project prior to construction starting on site, by submitting an F10 form if construction work is scheduled to last longer than 30 working days and has more than 20 workers working simultaneously, or if it exceeds 500 person days (Regulation 6).
- Taking reasonable steps to ensure that the principal designer is complying with their duties (Regulation 4), including ensuring that you comply with your designer duties (Regulation 11).

Regulation 7 deals specifically with the duties of domestic clients, which delegates the client duties to the principal designer and principal contractor. If you have a domestic client and they fail to make these appointments, the designer in control of the pre-construction phase of the project is the principal designer (Regulation 7(2)(a)) and the contractor in control of the construction phase of the project is the principal contractor (Regulation 7(2)(b)).

Use your client briefing process to assess your client's level of knowledge and experience of the CDM Regulations to decide the appropriate detail and format for the advice you need to give them regarding their duties. Do not wait until you have been formally appointed to do this because it may be too late, particularly if you are involved at the bid or feasibility stage of the project. If a principal designer has been appointed, request written confirmation that the client has been adequately advised of their duties to ensure your obligation as a designer is met.

Discuss design risk management with your client at the outset of your project. The greatest opportunities for positive influence on the safety of your project without design compromise are at the concept and pre-planning design stages of the project. The strategic decisions you make with your client regarding height, massing and spatial arrangements directly impact on how safely your project can be constructed, occupied, managed, maintained and refurbished and demolished. The design, quality and safety of your project may be undermined if residual risks you failed to design-out at the early design stages have to be controlled and managed by employing safety systems that may result in a greater likelihood of accidents.

When you prepare your design – or modify a design prepared by others – you have a duty to consider the general principles of prevention and any pre-construction information provided by your client to eliminate, so far as reasonably practicable, foreseeable risks to the health and safety of any person. This includes anyone involved in the construction, occupation, management, maintenance, refurbishment and demolition of your project.

If you are unable to eliminate any risks, after you have taken steps to reduce their impact through your design, you have a duty to provide your client's principal designer with clear, concise information regarding these residual risks. This may be in the form of a residual risk register, annotations or safety symbols on your model or drawings, or schedules on drawings (sometimes referred to as Safety, Health and Environment boxes, or SHE boxes).

If you have not designed-out all unusual or significant risks from your project, let the principal designer know, so that everyone is aware that special measures are required in connection with the architectural elements of the project. Do not be tempted to produce a generic document with common or insignificant risks that a competent contractor or building manager ought to already be aware of simply for the sake of being able to demonstrate that you have undertaken a risk assessment and produced a risk register. If you have successfully eliminated all unusual or significant risks advise your client and principal designer so they can advise the principal contractor accordingly.

In addition to residual risk information, you have a duty to provide information about the design, construction and maintenance of the architectural aspects of your project. This will assist your client, all other designers and any contractors to comply with their duties under the CDM Regulations.

6.3 Regulation 11 – Principal designer duties

If it is reasonably foreseeable that more than one contractor will be working on your project at any time, your client has a duty under the CDM Regulations to appoint a designer to be the principal designer with control over health and safety in the pre-construction phase of the project.

The principal designer's appointment must be in writing and made as soon as practicable before the construction phase begins, ideally at the outset of RIBA Work Stage 2 'Concept Design'. If your client fails to appoint a principal designer, they must fulfil the principal designer's duties. If your client is a domestic client, the designer in control of the pre-construction phase will be the principal designer, this will typically be the lead designer.

If you are appointed as principal designer, make sure your principal designer appointment is separate from your designer appointment, including a separate fee and schedule of services. Your statutory duties as a principal designer are separate and distinct from your statutory duties as a designer and as the principal designer duties do not include any actual design duties you should ensure your appointments are drafted accordingly to avoid creating additional contractual principal designer duties that extend your liability beyond that imposed by the regulations. You also need to ensure you have sufficient dedicated resources in place to provide your principal designer services appropriately.

Your principal designer appointment should be in place for as long as you are working for your client and there is a need for pre-construction design services. Bear in mind that the pre-construction phase design work for some elements of work may run simultaneously with work that has already reached the construction phase, for example it is not unusual for elements of the RIBA Work Stage 4 'Technical Design' to be carried out by designing sub-contractors and specialists at the same time as other elements of the works are being constructed on site. Accordingly, your appointment and principal designer duties will continue during the construction phase.

The principal designer must be a direct appointment with your client. If you are working on a design and build project and your design appointment is novated to the contractor, your principal designer appointment with your client may need to be terminated to avoid any post-novation conflicts of interest. In this case, your client will need to appoint a new principal designer, who may be the principal contractor, to replace you and to manage the coordination of the health and safety aspects of any remaining pre-construction information.

In addition to the general duties under Regulation 8, Regulation 11 sets out the duties of a principal designer. Your primary duty as a principal designer is to plan, manage and monitor the pre-construction phase to ensure that, so far as is reasonably practicable, the project is carried out without risks to health and safety.

The pre-construction phase covers any phase of your project during which design work is being carried out i.e. RIBA Work Stages 2 to 4. Bear in mind that some of the design you are completing during Work Stages 3 and 4 may continue after some sections or packages of work have commenced manufacture, assembly or construction at Work Stage 5, so it is not uncommon for your principal designer's duties in connection with the pre-construction to continue during the construction phase.

Whilst your duties as a principal designer do not include any specific design duties (these fall under your appointment as a designer), you still need to apply the principles of prevention and to identify, eliminate and control, so far as is reasonably practicable, foreseeable risks to anyone involved in the construction, occupation, management, maintenance, refurbishment and demolition of your project. This requires you to consider all available information when decisions are being taken by your client and the other designers regarding the planning and sequencing or phasing of design and construction work, including estimating the time and resources required to complete the work. This may include a need to remind your client of their duties in relation to managing the project and ensuring that enough time and resource are available to the design and construction teams.

The information likely to be available to you includes pre-construction information provided by your client, any construction phase plans that have been produced by the principal contractor and are relevant to design taking place during the construction phase, and any existing health and safety file provided in relation to the existing construction of your project.

Whilst your client is responsible for providing you with any pre-construction information that is already in their possession, as principal designer you will need to assist the client to collate this information, assess its adequacy, identify any gaps and advise the client regarding what additional information may be required. You are then responsible for ensuring the pre-construction information is provided in a convenient form to any designers and contractors working on your project – or being considered – to enable them to carry out their duties. Discuss and agree with your client the most efficient way to manage this process. Consider with your client whether they, or the principal contractor if appointed, should host an electronic data management system (EDMS) that is made available to the whole project team. This will enable you to ensure that all the relevant pre-construction information is available to the team as and when they need it, as well as providing you with the means to manage and maintain accurate digital records of who has received what information and when.

To fulfil your principal designer duties, you need to ensure that, so far as is reasonably practicable, all other designers working on your project comply with their duties under Regulation 9 and that everyone working on the pre-construction phase cooperates with your client, with you and with each other.

The most effective way to achieve this is to implement early, regular and effective communication between all the other designers working on your project (including specialists and contractors with design responsibility). The best way to do this is to ensure design risk management is incorporated into all regular design workshops or design team meetings and in progress meetings with the client and principal contractor. Use these regular meetings to discuss design risks, agree control measures required for risks that cannot be eliminated and agree the format and content of pre-construction information that the principal contractor will require to prepare the construction phase plan.

As the principal designer, you are not expected to be a health and safety expert or adviser, and you are not responsible for the other designers' designs or for advising the other designers on how they should eliminate risks or modify their design.[4] This remains each designer's responsibility under the CDM Regulations. You are responsible for ensuring that together with the other designers you identify and eliminate or control, so far as is reasonably practicable, the foreseeable risks in the design and coordinate this information to ensure clear, concise and relevant project specific information regarding significant and unusual residual risks is communicated to the client and principal contractor. Promptly alert your client if you have any concerns regarding a designer's skills, knowledge or experience, or the ability or willingness to carry out their duties appropriately. Remind your client, when you do so, that it is their duty to ensure that the designers they appoint are competent to undertake the design they have been appointed to deliver.

Regulation 12 sets out the principal designer's duties with respect to the construction phase plan and health and safety file.

As principal designer, you must assist the principal contractor in preparing the construction phase plan for your project by providing the principal contractor with a copy of all the relevant pre-construction information you have obtained from the client, and the relevant design and residual risk information you have obtained from the other designers. Again, you may find that utilising an EDMS may be the most efficient and effective way to do this.

During the pre-construction phase, and for the duration of your appointment as principal designer, you are responsible for compiling the health and safety file and you should commence work on this at the outset of your project. Collate contemporaneous and record copies of preliminary information, particularly at key stages of your project, so that your draft health and safety file provides an accurate record of your project at any point in time. Subsequently, regularly review and update it to take account of the progress of the works and any changes to the design and construction. Even if you know your appointment as principal designer will be terminated prior to commencement of the construction phase, you should prepare a health and safety file with the relevant pre-construction and designers' information available to you before your appointment is terminated.

The health and safety file should only include project specific information that is clear, concise and likely to be relevant to ensuring the safety of anyone planning, or carrying out, future maintenance or construction work on your project. As principal designer you are not responsible for producing this information. Your duty is to collate information provided by the other designers, including any designing contractors, sub-contractors and/or specialists. Remind the designers that they should not provide you with generic information, risk assessments or information regarding obvious hazards or risks that have been eliminated during the pre-construction or construction phases.

If you are appointed as principal designer for the duration of the construction phase of your project, pass the completed health and safety file to your client at the end of the project. If your appointment is terminated prior to completion of your project, pass the health and safety file to the principal contractor, who will assume responsibility for completing it and passing it on to the client when the project completes.

As principal designer, or in your capacity as a designer, you are not responsible for ensuring the other designers, the principal contractor or the other contractors have the necessary skills, knowledge and experience; this is your client's responsibility. The only exception is if you are appointing a designer as a sub-consultant. In this case you become their client and assume the client duties with respect to ensuring they have the relevant skills, knowledge and experience to fulfil their designer duties.

You are also not responsible for providing health and safety advice to your client or the project team, other than in respect of their duties under the CDM Regulations. If you, your client or any members of the design and construction team require specialist health and safety advice, recommend that your client appoints a health and safety adviser.

If you do not have the appropriate or relevant skills, knowledge and experience to provide principal designer services for your project (bearing in mind what is appropriate or relevant will vary from one project to another), one way to develop this experience is to appoint a health and safety adviser as your sub-consultant to provide you with the relevant advice to enable you to discharge your duties. Be aware, however, that you cannot delegate your statutory obligations under the CDM Regulations to a sub-consultant and that you remain liable under the regulations for discharging your duties correctly.

6.4 CDM Regulations and the Building Safety Act

The dutyholder regime under the CDM Regulations and the **Building Safety Act** (BSA), and more particularly **The Building (Appointment of Persons, Industry Competence and Dutyholders) (England) Regulations** (Dutyholder Regulations), are similar but not the same.

As outlined above, the dutyholder regime under the CDM Regulations applies to all construction projects carried out in Great Britain. These duties apply to construction projects as a whole and are intended to secure the health and safety of any person affected by the project, with a particular emphasis on the construction work, maintenance, cleaning and use of a project as a workplace.

The dutyholder regime under the BSA applies to all construction projects carried out in England, with enhanced duties in respect of higher-risk buildings. These duties apply to all projects involving construction work that is subject to building control approval and are intended to ensure buildings are designed and constructed in accordance with the relevant requirements of the Building Regulations.

Whilst the duties under the BSA have been created with the intention that they are aligned with those under the CDM Regulations they include some material differences that you need to be aware of before accepting an appointment to provide either designer or principal designer services, particularly if you intend to provide services in connection with both pieces of legislation.

As the scope of the BSA is broader than that of the CDM Regulations so are the competence requirements, which require you to have an appropriate level of knowledge of the relevant requirements of the building regulations. Depending on the nature of the project(s) you are working on this may require skills, knowledge and experience beyond that required to secure health and safety matters under the CDM Regulations.

Both the CDM Regulations and the BSA require you to have the necessary skills, knowledge and experience to undertake your role as a designer. The BSA however also requires you to have the necessary behaviours, essentially the ability to demonstrate that you are willing and able to make effective use of your skills, knowledge and experience. Whilst it may be argued that this is also an implied duty under the CDM Regulations, the BSA makes it an express duty and therefore potentially subject to a requirement to evidence the same.

Both the CDM Regulations and the BSA are applicable to a lesser extent to domestic projects, and both make provision for the duties of domestic clients to be undertaken by their principal designer and/or principal contractor.

CHAPTER 7:
THE BUILDING SAFETY ACT 2022 AND BUILDING SAFETY REGULATIONS

In response to Dame Judith Hackitt's 'Independent Review of Building Regulations and Fire Safety'[1] the UK Government committed to fundamentally reform the building safety system. The **Building Safety Act 2022** (BSA)[2] is the primary legislation that provides for this by giving effect to the policies set out in the government response to the Building a Safer Future consultation.[3]

Together with a number of statutory instruments, secondary legislation enacted under the BSA (see Chapter 5), the BSA strengthens the regulatory regime for building safety and acts as a vehicle for wider legislative amendments, including amendments to the **Architects Act 1997**,[4] the **Building Act 1984**,[5] the **Defective Premises Act 1972**[6] and the **Regulatory Reform (Fire Safety) Order 2005**.[7]

Together these legislative changes, which are applicable to all projects undertaken in England, are intended to create greater accountability and responsibility for fire and structural safety issues and Building Regulations compliance throughout the lifecycle of buildings, many of which have a direct impact on how we discharge our professional duties as designers.

In this chapter we consider:

7.1   Overview of the Building Safety Act
7.2   Role of the Building Safety Regulator
7.3   Building control reform for all buildings
7.4   Building control regime for higher-risk buildings

Consideration of all the provisions of the BSA is beyond the scope of this guide. The guidance in this chapter is limited to the aspects of the BSA that you need to be aware of and understand to enable you to discharge your duties as a designer appropriately. For more guidance regarding your duties as a principal designer under the BSA refer to the *RIBA Principal Designer's Guide*.

7.1 Overview of the Building Safety Act

The BSA gained royal assent on 28 April 2022 and is described as: 'An Act to make provision about the safety of people in or about buildings and the standard of buildings, to amend the Architects Act 1997, and to amend provisions about complaints made to a housing ombudsman.'

The BSA is arranged in six parts and has nine schedules:

- Part 1 provides an overview of the structure and provisions of the BSA that are intended to secure the safety of people in or about buildings and to improve the standard of buildings.
- Part 2 contains provisions about the Building Safety Regulator ('the Regulator') and its functions in relation to buildings in England.
- Part 3 sets out amendments to the **Building Act 1984**, including that the Regulator is the building control authority for higher-risk buildings in England.
- Part 4 sets out provisions for occupied higher-risk buildings in England.
- Part 5 sets out further provisions including powers to make provision about construction products, further provision about fire safety, and provision about the regulation of architects.
- Part 6 contains general provisions.

Of the nine schedules, the most relevant to us as architects and designers are:

- Schedule 1 – Amendments of the **Health and Safety at Work etc Act 1974**, which relate to the Regulator
- Schedule 3 – Cooperation and information sharing
- Schedule 11 – Construction products regulations.

As primary legislation, the BSA is an enabling act that makes provision for further legislation, including amendments to existing legislation, that contain the detailed duties relevant to our day-to-day design activities. There are a number of statutory instruments, secondary regulations, enacted under the BSA and collectively referred to in this guide as the building safety regulations. Those most relevant to designers include:

- **The Higher-Risk Buildings (Descriptions and Supplementary Provisions) Regulations 2023,** which came into force on 06 April 2023.
- **The Building (Approved Inspectors etc. and Review of Decisions) (England) Regulations 2023**, which came into force on 01 October 2023.
- **The Building Regulations etc. (Amendment) (England) Regulations 2023**, which came into force 01 October 2023.
- **The Building (Higher-Risk Buildings Procedures) (England) Regulations 2023** (HRB Regulations), which came into effect on 01 October 2023.

Together the BSA and the building safety regulations prescribe procedural requirements for the design, construction and building control approval of building projects in England, including refurbishment projects and those that include a change of use. The majority of these procedural requirements are not entirely new or unduly onerous management concepts. In many instances they simply reflect good practice with respect to project running, design coordination, document control and information management, which you are likely to already be familiar with and may already be employing to varying degrees in connection with the design of your projects, albeit on a professional rather statutory basis.

What is new is that the BSA and building safety regulations make it a legal duty to employ these management procedures on projects, with additional requirements for higher-risk buildings, to ensure that the design and construction of buildings is planned, managed and monitored appropriately and information regarding the same is available to all those that need it in a suitable format and at the time they need it.

The BSA and building safety regulations do not contain any requirements or guidance regarding the technical design or construction of buildings, other than by reference to the applicable or relevant requirements of the Building Regulations. Guidance regarding technical design and routes to compliance continues to be provided by statutory and non-statutory guidance, for example, the Approved Documents to the Building Regulations, as described in Chapter 5. The Regulator does however have overall responsibility under the BSA for regularly reviewing and updating relevant statutory guidance and under these powers will periodically review and, where appropriate, publish new and updated versions of the Building Regulations Approved Documents.

The overarching aim of the new regulatory regime is to ensure that everyone involved in the procurement, design and construction of projects takes responsibility for their part in the project. For designers, this means ensuring that you understand your client's duties as well as your own duties, to ensure that any design you are responsible for complies with the relevant requirements of the Buildings Regulations. To enable you to do this, it is important that you are aware of and understand all of your designer duties as set out in each of the various statutory instruments listed above, and how you ought to coordinate the discharge of these duties with those of the other dutyholders employed in connection with your project, including your client. The following guidance provides an overview of these duties and responsibilities, but it is important that you understand the details of the legislation as they relate to your projects. Ultimately, provided that you are diligent in the preparation of your design, including coordinating said design with others, and accept ownership and responsibility for your design in accordance with the spirit and intent of the building safety regulations, you will be on the right track to discharging your duties.

7.2 Role of the Building Safety Regulator

Part 2 of the BSA establishes the Health and Safety Executive (HSE) as the Building Safety Regulator (referred to as the Regulator or the BSR) with the responsibility and authority to oversee regulatory reform and continuing functions and performance of the building safety system, including the building control regime in England.

The Regulator has two statutory objectives in the way that it discharges its building functions (its duties under the BSA, the Building Act and HSWA).[8] These are to:

- secure the safety of people in or about buildings in relation to risks arising from buildings.
- improve the standard of buildings.

In meeting these objectives, the Regulator has an overarching duty to make sure that its building functions are carried out in a way that is transparent, accountable, proportionate and consistent, and that these building functions are targeted to only address issues where action is needed. The intention is that the Regulator's role is focused on improving the construction industry to secure safe outcomes on all projects without imposing unnecessary or wasteful administrative burdens on design and construction teams.

The three statutory building functions of the Regulator are:

- A duty to facilitate building safety for higher-risk buildings, which includes providing such assistance and encouragement to relevant persons, including designers, as it considers appropriate to secure the safety of people, in particular the safety of disabled people, in or about higher-risk buildings.[9]
- A duty to keep the safety of people in or about buildings and the standard of buildings under review, which applies to all buildings and not just higher-risk buildings.
- Facilitating improvement in the competence of industry and building inspectors by providing such assistance and encouragement as it considers appropriate to people working in the built environment industry,[10] including designers and registered building inspectors to improve their competence.[11]

Whilst the Regulator has the power to impose sanctions under the BSA, its building functions are primarily aimed at avoiding the need to rely on those sanctions. The intent of the building functions is primarily to support and encourage the construction industry to do the right thing in terms of designing safe and accessible buildings.

The Regulator also has the ability, following consultation, to propose new regulations under the BSA and a duty to establish a system for voluntary and mandatory occurrence reporting, which we consider in more detail in Section 7.4.[12]

To assist the Regulator to meet its objectives and fulfil its building functions it has a statutory duty to establish three committees:

- A **Building Advisory Committee** (BAC) to give advice and information to the Regulator about matters connected with any of the Regulator's building functions, except those in relation to competence. This committee replaces the Building Regulations Advisory Committee (BRAC) for England.[13]
- A **committee on industry competence** concerned with the competence of people, including designers, working in the built environment industry. This committee has prescribed functions including:
  - monitoring industry competence
  - advising the Regulator in relation to industry competence
  - advising people working in the built environment industry in relation to industry competence
  - facilitating people working in the built environment industry to improve industry competence
  - providing guidance to the public about ways of assessing the competence of people working in the built environment industry
  - carrying out analysis and research in connection with the above.
- A **resident's panel** of residents of higher-risk buildings, including residents who are disabled, to give the Regulator advice about matters in connection with its building functions and higher-risk buildings.

As well as seeking advice from the committees, the Regulator has the right to request assistance from local authorities and fire and rescue authorities (collectively referred to as 'the relevant authority') in connection with higher-risk buildings. This includes the power to direct a relevant authority to undertake a particular task within a specified period. This ensures that the Regulator is also able to seek the support of the relevant authority when they, the Regulator, have been appointed to act as the building control authority for non-higher-risk buildings (this is most likely to be relevant to mixed-use developments that include one or more higher-risk buildings and is intended to avoid the regulated functions for the site being split between the relevant authority and the Regulator). Schedule 3 of the BSA sets out specific reciprocal duties for relevant authorities and the Regulator to cooperate and share information in connection with their respective statutory functions. This is designed to foster a culture of joint working to ensure that the Regulator and the relevant authorities support one another to discharge their statutory functions effectively.

To ensure the effective functioning and decision making of the Regulator, the Regulator has enforcement powers under the BSA. This includes the power to authorise suitably qualified people to act as authorised officers to investigate building safety matters on the Regulator's behalf.[14] Authorised officers have the right to enter non-domestic buildings to investigate situations that are, or may be, dangerous; the right to take measurements and photographs, make recordings and take samples; and the right to seize material as evidence if it appears that there has been a breach of the BSA or the Building Act and there is a danger such evidence might otherwise be concealed, lost, altered or destroyed.

As designers, we have a duty to cooperate with the Regulator and its authorised officers, which includes a duty to provide information. The BSA makes it a criminal offence to provide false or misleading information, knowingly or recklessly, to the Regulator in connection with: any application made to the Regulator in connection with the BSA or Building Act; any purported compliance with the BSA or the Building Act; or for the purpose of avoiding enforcement action being taken. The Regulator will work with you to address any issues or concerns regarding the information you provide to avoid having to resort to imposing sanctions. However, if you fail to cooperate with the Regulator and are found guilty of providing false or misleading information you could be sentenced to imprisonment for up to two years and/or fined.

7.3 Building control reform for all buildings

The building safety regime implemented by the BSA, along with the building safety regulations, regulate and hold to account everyone participating in the design and construction of all new buildings, as well as the refurbishment of existing buildings. This is achieved by amendments to the parts of the **Building Act 1984** and **Building Regulations 2010** that deal with building control authorities and Building Regulations in England, including procedures for consulting with the Regulator regarding any regulations made under the BSA.

Provisions in the BSA, implemented as amendments to the Building Act and Building Regulations, that are pertinent to designers include:

- provision for dutyholders and general duties in respect to the Building Regulations
- competence requirements in respect to the Building Regulations (refer to Chapter 5)
- the role of building control authorities
- the procedural requirements of the Building Regulations
- regulation of the building control profession.

Providing guidance on all of the provisions – particularly many of the procedural requirements of the Building Regulations – is beyond the scope of this guide. However, some of the details that you need to be familiar with as a designer are set out below. As noted previously, these duties are not necessarily new, but make good practice in the design of buildings and, particularly in the case of higher-risk buildings, meet the minimum statutory requirement and are not just a 'nice to do'. The intention of these duties is to ensure that you take ownership and responsibility for all of the design work you undertake seriously – be it your own work or work procured from others under your control or employment – and that you understand and take responsibility for ensuring such work complies with the relevant requirements of the Building Regulations.

General duties, dutyholders and competence: The primary duties you need to be aware of and comply with as a designer are your general duty to plan, manage and monitor your design work and your duty to cooperate with others and share information for the purpose of ensuring compliance with the relevant requirements of the Building Regulations. These apply to all projects that you work on that are subject to building control approval, and not just higher-risk buildings.

Details of your duties are set out in the various statutory instruments with the primary duties set out in Part 2A of the Building Regulations (introduced by The Building Regulations etc. (Amendment) (England) Regulations). These impose general duties on you as a designer to take all reasonable steps to ensure the design work carried out by you (and anyone under your control) is planned, managed and monitored so that your design is such that, if the building work to which your design relates were built in accordance with your design, the building work would be in compliance with all relevant requirements of the Building Regulations. You also have a legal duty to cooperate with the client, other designers and contractors (including the principal designer and principal contractor, if there is one) to the extent necessary to again ensure that, if built, the building work to which your design relates is in compliance with all relevant requirements of the Building Regulations.

These general duties can be discharged by ensuring that you cooperate with your client and any other designers appointed by your client by engaging in regular communication, participating in regular client and design team meetings and regularly exchanging relevant information with the other designers regarding your design and its compliance, or non-compliance, with all relevant requirements of the Building Regulations. This includes identifying, discussing and resolving any problems or issues that may pose a risk to the compliance of your project with the relevant requirements of the Building Regulations so that these are addressed by you and the design team in a timely manner prior to submitting your project for building control approval and/or prior to construction.

As well as general duties, Part 2A of the Building Regulations includes duties specific to designers that state:

- You must not start design work unless you are satisfied that your client is aware of their statutory duties under the building safety regulations with respect to the building work you are designing.
- You must *take all reasonable steps* to ensure that, if built, the building work to which your design relates will be in compliance with all relevant requirements of the Building Regulations.
- You must *take all reasonable steps* to provide sufficient information about the elements of the design, construction and maintenance of the building you are responsible for to assist your client, other designers and contractors to comply with all relevant requirements of the Building Regulations.
- You must consider any design work completed by others that directly relates to the building work you are designing and report any concerns regarding the compliance of that design with the relevant requirements of the Building Regulations to the building regulations principal designer appointed by your client under Part 2A of the Building Regulations.
- If requested to do so, you must provide advice to the building regulations principal designer appointed by your client under Part 2A of the Building Regulations, or to your client, regarding whether the work relating to the design you are preparing or modifying is higher-risk building work.

When you are planning your services in connection with your designer duties under Part 2A of the Building Regulations, including planning your fees and resources, it is important you consider the standard of care required under the building safety regulations. A number of the designer duties are absolute obligations i.e. you *must* do something or you *need to ensure* a particular outcome. Failing to fulfil these obligations, which may require you to commit more management time or resources to your project, could result in a breach of your statutory duties under the building safety regulations. Other duties are qualified to *taking all reasonable steps* to achieve the intended outcome. Whilst this is a lower standard of care it still requires that you not only commit sufficient time and resources to plan and manage your services but also, and just as importantly, produce and manage the evidence required to demonstrate that you have discharged your duties and achieved the intended safe outcomes on your project by evidencing that all the design completed under your control meets all relevant requirements of the Building Regulations. For example, making sure you have sufficient documentary evidence that you have discharged your duties, such as copies of design reviews, written communication with the designers you have coordinated your design with, information management records and trackers, project reports, design team and client meeting minutes and site inspection reports.

An important point to remember regarding your statutory design duties under the building safety regulations is that you are and remain responsible for all the design work completed under your control, irrespective of any review, inspection or approval of your design, or the construction work to which it relates, by the relevant building control authority, including the Regulator if you are designing a higher-risk building. It is your responsibility to ensure that your design complies with the applicable or relevant requirements of the Building Regulations, and you need to be able to demonstrate to the relevant building control authority that this is the case by providing evidence of the same. Whilst approval by the relevant building control authority will verify that your design is compliant based on the documentary evidence you submit for building control approval, this does not relieve you of responsibility for any errors or omissions in your design. The building control authority may advise you of errors or deficiencies in your building control approval application or request additional information to support your assertion that a design complies with the relevant requirements of the Building Regulations, but it is not the building control authority's responsibility to check or interrogate your design. The building control authority does not assume any responsibility for the compliance of your design, or lack of it, with the applicable or relevant requirements of the Building Regulations.

We have already considered in Chapter 5 the general requirements for competence under Part 2A of the Building Regulations, including those requirements specific to designers, but it is worth noting here that your client, or the principal contractor if your client is a domestic client, has a duty under Part 2A of the Building Regulations to take all reasonable steps to satisfy themselves that you fulfil your competence requirements, or have arrangements in place for adequate supervision if you are in training, before permitting you to carry out any design work. If you are working on a higher-risk building project your client will also need to submit evidence of your competence to the Regulator as part of the building control approval application. To enable your client to do this it is important that you maintain appropriate details of your competence, including records of any training and continuous professional development undertaken that is relevant to your role, responsibilities and project.

The role of building control authorities: With respect to the role of building control authorities in England, the BSA makes it mandatory for the Regulator to be the building control authority for any design work you undertake on higher-risk buildings, or those that will become higher-risk buildings because of your proposed design. We consider in section 7.4 below the definition of higher-risk buildings.

Where the Regulator is not the building control authority for a project the local authority will be the building control authority.[15] In this instance, approval from the relevant authority for the commencement of construction may be sought by either the submission and approval of a building notice (previously known as an initial notice prior to the introduction of the Building Regulations amendments) or the submission of an application for building control approval (previously known as the deposit of plans prior to the introduction of the Building Regulations amendments). In this case, the building control functions (review of plans, site inspections and issue of notices) may be undertaken by either the local authority, or by a **registered building control approver** (RBCA), in both instances based on advice provided by a **registered building inspector** (RBI). Registered building inspectors and registered building control approvers are competent individuals registered by the Regulator that fulfil a function similar to that which approved inspectors undertook prior to the introduction of the building safety regulations. We consider the roles of registered building inspectors and registered building control approvers in more detail below.

Procedural requirements of the Building Regulations: Amendments introduced under the building safety regulations enable the Building Regulations in England to be used to set certain provisions relating to the procedures for applying for building control approval. These include:

- general procedures that may or must be followed in relation to any construction work, including the making of applications and the issue of notices and certificates.
- procedures relating to applications for building control approval, including building control authorities' power to impose requirements in connection with approvals and to approve changes to anything that has previously been approved.
- requirements for obtaining, keeping and giving information and documents, including prescribed standards for information and duties to keep information up to date.
- requirements for establishing a system of voluntary and mandatory occurrence reporting.
- requirements for the form and content of documents and information that must accompany an application for building control approval, including the way in which information is to be given to the building control authority.
- provision for the inspection and testing of work, buildings, services, fittings and equipment, including the taking of samples.
- power to extend, by agreement, the period a building control authority has to consider an application for building control approval.

As well as these general procedural provisions, other amendments to the Building Regulations made by the building safety regulations that you ought to be familiar with include:

- The automatic lapse of building control approval for work that is not commenced within three years of the date the application for building control approval was made, or the building notice or certificate was issued. If your project includes more than one building, approval will lapse for any or all of the buildings that are not commenced within three years, even if work elsewhere on the site has commenced.

Under the building safety regulations, commencement of work in relation to the construction of complex buildings (defined below) is completion of the sub-structure of a building up to and including the foundations supporting the building and the structure of the lowest floor level of the building (but not the other buildings or structures to be supported by those foundations).

For buildings that are not complex buildings, including extensions to existing buildings, work is regarded as commenced when the sub-surface structure of the building or the extension including all foundations, any basement level (if any) and the structure of the ground floor level is completed. For all other building work, commencement is deemed to be reached when 15% of the proposed work is completed.[16] To ensure there is no confusion or misunderstanding regarding when works have commenced and to avoid building control approval automatically lapsing, you should advise your client that they should submit a statement, or statements if there is more than one building in the development, to the relevant authority confirming the date when the works were sufficiently complete in accordance with the relevant definition of commencement.

- The right of the building control authority to issue compliance and stop notices where there is or is likely to be a contravention of the relevant requirements of the Building Regulations. Compliance notices are intended for use where there is, or is likely to be, a contravention of a non-safety related obligation under the Building Regulations, for example, errors or omissions in your design that may result in level access to, or within your building being compromised. Stop notices are intended for use where there is, or is likely to be, a contravention of the Building Regulations, which would present a risk of serious harm to people in or about the building, for example, inadequate or inappropriate design and specification of fire doors. Failure to comply with either a compliance or stop notice is a criminal offence with a penalty of an unlimited fine and/or two years' imprisonment.
- Details regarding the contravention of the Building Regulations, including making contravention of the Building Regulations, and the requirements imposed under the Building Regulations, a criminal offence subject to a penalty of an unlimited fine and/or two year's imprisonment. The time limit during which the building control authority

can enforce rectification in respect of a contravention of Building Regulations on your project is 10 years from the date of completion.

Complex building means:
(a) a building which is to be constructed on the same foundation plinth or podium as any other buildings or structure;
(b) a building which has more than one storey below ground level;
(c) a building where its proposed use is primarily as a public building where the public or a section of the public has access to the building (whether or not on payment) provided that the building has a capacity for 100 or more visitors.

Public building means:
(a) a shop or shopping centre;
(b) premises where food or drink are sold for consumption on the premises, including a nightclub, social club or dance hall;
(c) a stadium, theatre, cinema, concert hall;
(d) a sports ground;
(e) an exhibition hall or conference centre;
(f) a hospital or premises for the provision of health care.

The intention of these amendments is to ensure that all construction projects in England are designed and constructed to the current functional requirements of the Building Regulations to achieve the intended safe outcomes. Whilst the Regulator has the power to impose sanctions to achieve this, the purpose of the contravention notices is to avoid the need to rely on these sanctions. As we considered above, the emphasis of the building safety regulations is on your duty to communicate, coordinate, collaborate and cooperate with the rest of the project team to avoid harm, ensuring that you take responsibility for the compliance of your design from the outset to achieve positive outcomes, rather than simply aiming to satisfy the bare minimum required by regulation or relying on the building control authority to advise you how to complete your design.

Regulation of the building control profession: As we considered above, one of the Regulator's objectives is to improve competence and accountability in the building control sector. To achieve this, the building safety regulations create a unified professional and regulatory structure for building control by amending the Building Act to provide for the registration of building inspectors and building control approvers, which replace approved inspectors.

Registered building inspectors are individuals that the Regulator is satisfied are competent to provide advice to building control authorities or registered building control approvers and to carry out restricted activities, such as carrying out site inspections.

Registered building control approvers are individuals or organisations that the Regulator is satisfied are competent to carry out specified building control functions in relation to a building project, such as approving plans or submitting a building notice. A registered building control approver is required to obtain and consider advice from a registered building inspector before exercising a restricted function.

The Regulator is responsible for determining and publishing the criteria required for a building inspector, or building control approver, to gain entry to the register. These make provision for different classes of building inspectors that reflect the nature and extent of an individual's competence, qualifications or experience. Entry on the register for inspectors and approvers is for a prescribed period and subject to a code of professional conduct and practice for inspectors, and professional conduct rules for approvers, with sanctions for professional misconduct or contravention of the rules respectively.

A competent individual can apply to the Regulator to be: a registered building inspector to provide advice to others; a registered building control approver to undertake building control work under Part 2 of the Building Act, or; both, which would allow them to undertake Part 2 work and rely on their own expert advice. For all projects, your client can appoint an appropriately registered building inspector as part of the project team to provide you with advice regarding design compliance with the relevant requirements of the Building Regulations. For projects that do not include higher-risk buildings, your client can subsequently seek building control approval from a local authority building control or a registered building control approver, but if your client appoints a registered building control approver, this may not be the same individual or organisation as the registered building inspector already advising on your project. This is intended to ensure that the building control approval process remains independent of the design and construction process of your project, avoiding any potential conflicts of interest that could otherwise undermine the robustness of the regulatory regime.

The Regulator is responsible for maintaining the registers of building inspectors and building control approvers and making these publicly available, including prescribed details regarding each registrant's entry, which will enable you and/or your client to determine whether they have the competence appropriate to your project.

7.4 Building control regime for higher-risk buildings

Section 120D of the Building Act as amended by the BSA, defines a higher-risk building in England as a building that is at least 18m in height *or* has at least 7 storeys, and contains at least two residential units, a care home, or a hospital.[17] Buildings are not higher-risk buildings if they are comprised entirely of a secure residential institution, a hotel (including short-term holiday lets but excluding serviced apartments), or military premises.[18]

The methods for determining the height of a building and the number of storeys for the purposes of establishing whether a building falls under the definition of a higher-risk building are set out in **The Higher-Risk Buildings (Descriptions and Supplementary Provisions) Regulations 2023**.[19,20]

Determining the height of a building. The height of a building is to be measured from ground level to the top floor surface of the top storey of the building (ignoring any storey that is a roof-top machinery or roof-top plant area or consists exclusively of roof-top machinery or roof-top plant rooms).

Where the top storey is not directly above the lowest part of the surface of the ground adjacent to the building, the height of the building is to be measured vertically from the lowest part of the surface of the ground adjacent to the building to the point that is a horizontal projection from the top of the floor surface of the top storey of the building (ignoring any storey that is a roof-top machinery or roof-top plant area, or consists exclusively of roof-top machinery or roof-top plant rooms).

Storeys. When determining the number of storeys a building has the following is to be ignored:
(a) any storey which is below ground level;
(b) any storey which is a roof-top machinery or roof-top plant area or consists exclusively of roof-top machinery or roof-top plant rooms; and
(c) any storey consisting of a gallery with an internal floor area that is less than 50% of the internal floor area of the largest storey vertically above or below it which is not below ground level.

Where there is more than one building on your project, any storey directly beneath the building(s) that is not below ground level is to be counted in determining the number of storeys the building has. For example, a sitewide podium that is above ground level.

A storey is treated as below ground level if any part of the finished surface of the ceiling of the storey is below the ground level immediately adjacent to that part of the building.

The BSA gives the Secretary of State the power, by regulations and based on advice from the Regulator, to amend or supplement the definition of higher-risk buildings by reference to a building's size, design, use or any other characteristic. Any amendment to the definition is conditional on the Regulator confirming that it is required to manage a building safety risk, which could otherwise cause serious injury or death to a significant number of people. You need to ensure that you keep yourself informed of any consultation or proposed changes to the regulations, including the definition of higher-risk buildings, as they apply to your projects. This will enable you to advise your client as to whether their project is, or may become during the course of its design and/or construction, a higher-risk building.

All design, construction and refurbishment work on higher-risk buildings in England is regulated by the Regulator and is deemed to be 'higher-risk building work'.[21] Therefore, the Regulator is the default building control authority for your project if your project includes higher-risk building work. In this case, the Regulator is responsible for supervising and enforcing compliance with all the applicable requirements of the Building Regulations for your project, not just fire and structural safety matters. As the Regulator is the building control authority, the local authority is no longer responsible for acting as the building control authority, or for enforcing the applicable requirements of the Building Regulations.

The Regulator will also become the building control authority for your project if it includes a non-higher-risk building that becomes one as a result of the proposed building work – for example, if you propose a change of use or add an additional storey so that your building falls within the definition of higher-risk building – or where a higher-risk building ceases to be one as a result of the proposed building work.

Your client may also choose to have the Regulator as their building control authority for a non-higher-risk building, subject to your client gaining the Regulator's agreement and submitting a *regulator's notice* to the local authority advising the local authority of the same. This is most likely to be appropriate if your client's development plans include a mix of higher-risk and non-higher-risk buildings and your client wants to avoid having different parts of the project being supervised by the Regulator and the local authority. Your client can only issue a regulator's notice if neither a building control approval application nor a building notice has already been made or issued to the local authority (including via a registered building control approver) in connection with the relevant part(s) of the project.

The details of the building control regulatory regime for higher-risk buildings under the BSA and amended Building Act, are set out in the *HRB Regulations*.

This is a significant piece of legislation that has seven parts and three schedules:

- Part 1 provides preliminary details, including definitions to assist interpretation of the regulations.
- Part 2 sets out the building control approval procedures for new and existing higher-risk building work (HRB work).
- Part 3 sets out the statutory regime for managing changes proposed to the approved design before or during construction.
- Part 4 sets out duties in relation to the golden thread, mandatory occurrence reporting system and handover of information on completion.
- Part 5 sets out the application procedures for completion and partial completion certificates.

- Part 6 sets out the regime for the Regulator's site inspections and the review and appeal of the Regulator's decisions.
- Part 7 deals with a number of miscellaneous items, including the Regulator's right to direct that building control application submissions, directions and documentation are dealt with electronically.
- Schedule 1 prescribes details regarding the documents required to accompany a building control approval application for HRB work.
- Schedule 2 defines the work that is exempt from the HRB Regulations.
- Schedule 3 outlines the transitional provisions that were relevant to projects that were sufficiently progressed prior to 06 April 2023 and were exempted from the building control regime for HRB work.

A detailed analysis of all parts of the HRB Regulations lies outside the scope of this book, but the details that you need to be familiar with if you are involved in the design of HRB work include:

- building control approval application regime for HRB work (Gateway 2)
- change control process for HRB work
- golden thread and key building information
- mandatory occurrence reporting
- completion certification application process for HRB work (Gateway 3)
- prescribed documents for HRB work applications.

As noted above, these regulatory requirements deal with the procedural aspects of preparing and submitting a building control approval application for HRB work to the Regulator. They do not contain any detail or guidance regarding the technical aspects of your design or routes to compliance with the Building Regulations. Responsibility for identifying the relevant technical requirements of, and demonstrating compliance with, the Building Regulations remains with you and the rest of the design team. The procedures establish the minimum scope and nature of the design information required and how it should be submitted to support your building control approval application. Whilst the scope and format of the documents is prescribed, none of the information required should pose a particular challenge to a competent design and construction team, provided that you and your client's project team have agreed procedures in place to plan, manage and coordinate your design information from the outset of your project.

Building control approval regime for HRB work (Gateway 2): Building control approval for HRB work must be granted by the Regulator before any HRB work, or a stage of HRB work, commences, which includes work on an existing higher-risk building. This point in the approval process is referred to as Gateway 2.[22] It is deemed to be a 'hard stop' because the construction of any building work cannot legally proceed unless or until the Regulator has granted approval for that work. Building work is defined in Part 2 of the Building Regulations and is limited to new construction excluding site investigation, survey and preparation, demolition and strip-out work, which can all proceed prior to gaining Regulator approval for the HRB work.

A building control approval application for HRB work must be submitted electronically by your client, or by another person acting on your client's behalf, for example, the building regulations principal designer. If the application is submitted by a person acting on your client's behalf, the application must be accompanied by a statement signed by your client confirming your client agrees to the application being made and that the information contained in the application is correct. The application must be made in writing and include prescribed information regarding the proposed works and be accompanied by specific project information, all in accordance with Regulation 4 of the HRB Regulations.

As well as information regarding the design and specification of the HRB work, Regulation 4 also requires information regarding the procurement and programming of the work to be submitted for approval. As compiling this information will require input from the rest of your client's project team, it is important to agree with your client who will plan, manage and monitor the building control approval application and who will be responsible for preparing each of the documents required, bearing in mind that it is likely to require potentially significant time and resources to manage the collation and coordination of information from several members of the project team. This responsibility can be delegated to the lead designer or building regulations principal designer, but this is not a statutory requirement.

If you are appointed by your client to prepare the building control approval application, it is sensible for you to use the details set out in Regulation 4 as a checklist for verifying you have received everything you need from the client and the rest of the project team to ensure that your application is complete before submitting it to the Regulator for approval. A copy of Regulation 4 is provided in Appendix II.

The Regulation 4 building control approval application information requirements include:

- Client, project team and project details, including specific information regarding the site and a description of the proposed HRB work.
- A **competence declaration**, which is a statement that contains a declaration in relation to the principal designer and principal contractor responsible for the design and construction of the HRB work, a stage of HRB work or work to an existing higher-risk building. In relation to the principal designer (or sole or lead designer) the declaration must state that the client, having taken all reasonable steps and subject to a declaration of truth, is satisfied that the principal designer fulfils the serious sanctions regarding past misconduct and competence requirements set out in Part 2A of the Building Regulations.
- A **construction control plan**, which should set out the strategies, policies and procedures your client has adopted to:
  - plan, manage and monitor the HRB work to ensure compliance with the applicable requirements of the Building Regulations, including arrangements to record evidence of compliance and maintain the golden thread information.
  - identify, assess and keep under review the competence of persons involved in the design and construction of the HRB work, including procedures to be followed in respect of any serious sanctions and consideration of past behaviour which might call into question the suitability of a person to be appointed.

- support cooperation between the designers, contractors and others involved in the HRB work, including the sharing of all necessary information.
- maintain a schedule of all the appointments your client has been made at the date of the application, including details of the principal designer (or sole or lead designer) and principal contractor and any appointments they have made, and a summary of each appointee's responsibilities.
- regularly review the construction control plan.

- A **change control plan**, which must set out the strategies, policies and procedures the client has adopted to ensure any controlled changes are recorded and take place in accordance with Part 3 of the HRB Regulations.

- A **mandatory occurrence reporting plan**, which sets out the steps taken by the principal dutyholders to establish a mandatory reporting system in accordance with Part 4 of the HRB Regulations before the construction phase begins.

- A **Building Regulations compliance statement**, which must set out the approach taken to designing the proposed higher-risk building or designing work to be carried out on a higher-risk building, including the building standards you have adopted in the design. This document must include details of any assumptions you, and other designers, have made in the course of your design regarding the characteristics and behaviours of the intended occupants of your building, including justification for these assumptions and the proposed management and maintenance strategy for the building. This document must also include details of the approach you have taken to ensure compliance with all applicable requirements of the Building Regulations and, if relevant, reasons for adopting a route to compliance that does not follow the guidance in the relevant Building Regulations Approved Documents. This should include why you consider that the alternative approach is appropriate and explain how it ensures compliance with all applicable requirements of the Building Regulations.

- A **fire and emergency file**, which must explain the approach to identifying, assessing and managing building safety risks to ensure compliance with the applicable requirements of the Building

Regulations relating to building safety risks.[23] Again, this should include details of any assumptions you have made regarding the intended occupiers of the building and how this might impact on the management of building safety risks.
- A **partial completion strategy**, which is required if your client intends to occupy part or parts of your project before all of the HRB work is complete. This strategy must explain the design strategy you have adopted to enable the safe occupation of each part of your project as it is completed and the proposals for managing and maintaining each part of the building to be occupied, including any assumptions made as to the intended occupiers in respect of the intended management or maintenance strategies.

Details regarding the information that must be included in each of these documents is provided in Schedule 1 'Documents' of the HRB Regulations. A copy of Schedule 1 is provided in Appendix III.

Bearing in mind the comprehensive nature of the information required, it is sensible for you and your client to engage with the Regulator prior to submitting an application to discuss your project and the documents you intend to submit for approval. You should also consider giving the Regulator at least two weeks' advance warning before submitting your application so that they can be ready to start work on it as soon as you submit it.

Once the Regulator has received your building control approval application the Regulator must determine whether the application is valid and notify your client, or the applicant acting on your client's behalf, accordingly. A valid application will proceed to determination. Where the Regulator considers that an application is not valid the Regulator must include reasons why in the notification to your client.

Following validation of your building control approval application, the Regulator has twelve weeks to determine whether to approve or reject it. If the Regulator requires a longer period to consider your application, the Regulator and your client may agree in writing an extension to this statutory approval period. The Regulator has a statutory duty to consult regarding your application with the relevant fire and rescue authority with respect to structure, fire safety and access (Parts A, B and M of Schedule 1 to the Building Regulations) and, if relevant, the sewerage undertaker with respect to Part H4 of Schedule 1 to the Building Regulations.[24] Such statutory consultees have fifteen working days to review your proposed plans during which time the Regulator is prevented from approving the application.

If the Regulator is unable to reach a decision regarding your application within twelve weeks and does not agree an extension with your client, your client is entitled to apply to the Secretary of State to determine the application on the Regulator's behalf.[25] In the absence of an extension of time or application to the Secretary of State, if the Regulator does not approve your application within twelve weeks it is deemed to be refused and commencing construction of the HRB work is a criminal offence.

This regime ensures that work not subject to appropriate regulatory oversight is not approved by default and enforces the 'hard stop' at Gateway 2 preventing construction starting prior to approval by the Regulator. This could have programme implications for your project, which you should discuss with your client at the outset to assess the impact of the approval period on the progress of your project and the risk of delays to the programme due to a protracted approval period or a rejection for non-determination.

If the Regulator does grant Gateway 2 approval for your building control approval application, they may also impose pre-commencement conditions, including the submission of further design details, documents or information for approval. Following receipt of the Regulator's approval, approved work can only commence once any pre-commencement requirements imposed by the Regulator have been fulfilled and notice of your client's intention to start work and the date that the work is due to start has been submitted to the Regulator, which needs to be submitted to the Regulator at least five working days before work commences.[26]

Not more than five working days after the day on which the HRB work has commenced your client must give notice to the Regulator to that effect. The definition of commencement in this instance is the same as that used for determining whether building control approval has lapsed which we considered above, so this is not the day that work started on site, but the date when the works to ground floor level or 15% of the work is complete, depending on the complexity and nature of the HRB work.

Change control process for HRB work: Part 3 of the HRB Regulations makes provision for *controlled changes* to be made to the approved design and construction of HRB work before or during construction. Any controlled change you, or another designer (including a designing contractor or sub-contractor), propose to the approved design of the HRB work is subject to a change control application that must be submitted for approval by the Regulator prior to the work affected by the change being carried out.

As well as variations to the approved design for the HRB work, a controlled change includes a change to any stage of HRB work, including adding or removing a stage, and the strategies, policies and procedures described in any of the current agreed building control application documents. Before any controlled change can be carried out, your client must ensure that a record of the controlled change is made in a change control log and a revised version of any agreed document affected by the change is produced. There are three types of controlled change: notifiable, major and recordable change.

Notifiable change means:
- a change to the construction control plan other than a change to the schedule of appointments contained within it;
- a change to the change control plan;
- a change to the layout of a flat or a residential room in a proposed higher-risk building or in a higher-risk building;

- a change to the number or dimensions of any openings in any wall, ceiling or other building element for any pipe, duct or cable;
- a change to the wall tie, wall restraint fixing or support system in any wall or proposed wall (excluding an external wall);
- a change of any construction product or building element to be used in or on a proposed higher-risk building (or to be used as part of works to a higher-risk building) where its replacement is of the same or higher classification under the reaction to fire classification (within the meaning in regulation 2(6) of the 2010 Regulations) (except where an agreed document specifies the use of a construction product or building element falling within a specified class, the change to another product or element falling within the same design specification is a recordable change);
- any other change to the fire and emergency file or the fire safety compliance information;
- a change specified by the regulator as a notifiable change by notice given in accordance with regulation 25 (change control: regulator power to specify notifiable changes and major changes);
- a change to the partial completion strategy;
- a change to a staged work statement or a subsequent stages statement.

Major change means:[27]
- a change which is a material change of use to any part of a proposed higher-risk building;
- a change of the proposed use of any part of a proposed higher-risk building so that after the change the part of the building is to have a use as a car park or cease to have a use as a car park (whether the car park is for the occupiers of the building or otherwise);
- a change which increases or decreases the external height or width of a proposed higher-risk building;

- a change to the number of storeys in a proposed higher-risk building (including adding or removing a mezzanine or gallery floor);
- a change to the structural design or structural loading of the building;
- a change to the number or width of the staircases in a proposed higher-risk building or a change to the length of any other escape route or the number or width of any escape route (including evacuation lift) within the proposed building;
- a change to the external wall of a proposed higher-risk building including a wall tie, wall restraint fixing or support system in the wall;
- a change to any part of the active fire safety measures or passive fire safety measures in a proposed higher-risk building referred to in the fire and emergency file;
- a change to the layout or dimensions of the common parts within a proposed higher-risk building;
- a change of any construction product or building element to be used in or on a proposed higher-risk building such that its replacement is of a lower classification under the reaction to fire classification (within the meaning in regulation 2(6) of the 2010 Regulations(18)) (except where an agreed document specifies the use of a construction product or building element falling within a specified class, the change to another product or element falling within the same design specification is a recordable change);
- a change to any assumptions made in the design of a proposed higher-risk building as set out in the Building Regulations compliance statement;
- a change proposing occupation of any part of the proposed higher-risk building before all the work is completed;
- a change to the number of flats, residential rooms or commercial units contained in a proposed higher-risk building.

> **Recordable change** means a controlled change which is neither a major change nor a notifiable change.

Where two or more controlled changes are related to the same change to the building work, these are referred to as *related changes*. If at least one of these controlled changes is a major change and one is a notifiable change, the changes should be included in the same change control application.

Where a change control application has been submitted in relation to a proposed controlled change, which is referred to as the *main change*, and it is proposed that another controlled change (excluding recordable changes) is required as a consequence of the main change, this change is referred to as a *consequential change*. A notification or change control application for a consequential change cannot be submitted to the Regulator until the Regulator has approved the change control application for the main change.

Work deemed to be a notifiable change cannot commence on site until a notice of the controlled change has been submitted to the Regulator. The notice must include details of the proposed controlled change, why the change has been proposed, whose advice has been sought regarding the change and a summary of the advice provided, an assessment of the agree documents affected by the proposed change and a *compliance explanation* in relation to the proposed change.

> **Compliance explanation** means an explanation, in relation to a proposed change, of how:
> (a) the HRB work, the stage of HRB work or work to existing HRB will, after the proposed change is carried out, comply with all applicable building regulations; and

(b) the strategies, policies and procedures in any agree document (including in relation to controlled changes, mandatory occurrence reporting, competence of persons, or sharing of information and cooperation) will, after the proposed change is carried out, comply with the relevant requirements of the Building Regulations, including the provisions for changes to documents, the golden thread, mandatory occurrence reporting and Part 2A of the Building Regulations.

A change deemed to be a major change cannot commence on site until the Regulator has approved a change control application for the same. Change control applications for major changes are subject to the same statutory provisions regarding validity and consultation as a building control approval application, including consultation regarding fire or sewerage. The Regulator has six weeks from the date of receipt of a major change application to determine the application, or longer if agreed in writing with your client. The Regulator's approval of a controlled change may include pre-commencement conditions including the submission of further information or documents for approval prior to the commencement of work relating to the controlled change. Notice of the Regulator's rejection of a change control application must include reasons for the rejection.

The Regulator has the power to determine that any proposed controlled change is a notifiable or major change, in relation to specified work, by giving notice in writing to that effect to the client, principal contractor and principal designer.[28] To avoid the risk of proceeding with abortive work in relation to a controlled change, particularly where it is not clear whether a change may be a notifiable or major change, it is worth discussing this with the Regulator prior to submitting a notice or application for a controlled change.

Details of every controlled change must be recorded in a **change control log**, which is a document that must be created and maintained by the principal contractor (or sole contractor) for the purpose of recording prescribed information in respect of all changes to your project.[29]

As well as controlled changes to the design of the HRB work, the Regulator must be notified if, at any time after building control approval for HRB work is granted, there are changes to your client, the principal designer or principal contractor. Where there is a change in client, your new client must be provided with a copy of the golden thread information, a copy of the project information and a document explaining the arrangements made under Part 2A of the Building Regulations.

Golden thread: Your client has a legal duty to make arrangements for an electronic facility to be created and maintained by them, or someone acting on their behalf, for the purpose of holding a digital copy of your project's *golden thread information*.[30] As a designer of HRB work you have a legal duty to provide copies of any design you are responsible for preparing as part of the building control approval application. These must be provided to your client's building regulations principal designer for inclusion in the golden thread information prior to the construction phase beginning.

The golden thread information for HRB work includes as a minimum:

- details of your client's arrangements for maintaining the golden thread information.
- a copy of the fire statement in relation to the HRB work.
- a copy of the plans and documents approved by the Regulator (including any revised documents approved under the change control procedures or agreed documents provided or approved to meet a Regulator's requirement) together with all evidence recorded to show compliance with the applicable requirements of the Building Regulations.
- any notices and statements in connection with the replacement of the building regulations principal designer.
- a copy of any mandatory occurrence reports submitted to the Regulator.

- a copy of the completion certificate application (or partial completion certificate application) together with a copy of each document that will accompany the application.

The golden thread information must be kept in a standardised electronic format that is, and remains, transferrable and accessible to those who may need to access it, including the occupiers of the completed building, ensuring that the information does not get lost or corrupted when it is transferred. There is no prescribed format for the golden thread information, and it may be suitable for some projects to include it as part of a BIM model, but this is not required by the HRB Regulations. Whatever format is used, it is essential that only accurate and up to date information is held in the golden thread, that as far as reasonably practicable consistent language terminology and definitions are used throughout to describe the golden thread information and that it is in a readable format which is intelligible to all those that are intended recipients of it in the future, including residents of the completed project.

Future and ongoing access to the golden thread information by residents is an important consideration when determining the suitability of and access to the proposed digital information platform, this is particularly important if BIM is being considered, which may not be readily accessible to many residents.

The client's proposed arrangements for the golden thread information must be made available as soon as reasonably practicable following a request from the principal dutyholders (the building regulations principal designer and principal contractor) and must ensure that the golden thread information is secure from unauthorised access and is only changed in accordance with procedures that record who made the change and when.

No later than the date of completion of the HRB work, your client must give a copy of the golden thread information to the principal accountable person (PAP), an organisation or individual who owns or is appointed by the owner to be responsible for managing and maintaining the higher-risk building. The golden thread should be provided to the principal accountable person in the same format as it was maintained, retaining the filing structure and any information logically associated with it and providing any key needed to understand the data.

Mandatory occurrence reporting: Under the HRB Regulations, the principal dutyholders (the building regulations principal designer and principal contractor) must establish a system that enables, as far as is reasonably practicable, the prompt reporting of every safety occurrence to the dutyholders by any designer (including the principal designer), any contractor (including the principal contractor) and any other person who is a periodic visitor to the site of HRB work.

Safety occurrence means:
(a) in relation to a design, an aspect of the design relating to the structural integrity or fire safety of a higher-risk building that would, if built, meet the risk condition;
(b) otherwise, an incident or situation relating to the structural integrity or fire safety of a higher-risk building that meets the risk condition.

The **risk condition** is that use of the building in question without the incident or situation being remedied would be likely to present a risk of a significant number of deaths, or serious injury to a significant number of people.

As a designer you have a duty to report to the building regulations principal designer any safety occurrences that you are aware of. To enable you to do this, the building regulations principal designer must take reasonable steps to ensure that you are provided with adequate instruction and information on the mandatory reporting system set-up for your project before you start any design work. This includes giving you information on the incidents or situations that you should be reporting.

If you are the sole or lead designer for your project, you must ensure that site inspections of HRB design work are carried out at an appropriate frequency to identify any safety occurrences throughout the construction phase. This may be a system of regular inspections undertaken by you, or by another competent designer or consultant. Either way, your client and the principal dutyholders are responsible for establishing the inspection regime and seeing that it is maintained for the duration of the construction phase to ensure that any safety occurrences are identified and reported in accordance with the safety occurrence reporting system your client has established on your project.

The principal dutyholders are responsible for notifying the Regulator of any reported safety occurrence. Further guidance regarding the principal dutyholders duties in connection with mandatory occurrence reporting for higher-risk buildings is provided in the *RIBA Principal Designer's Guide*.

Completion certification process for HRB work (Gateway 3): Following completion of any HRB work, your client must submit a completion certificate application to the Regulator for approval in accordance with Part 5 of the HRB Regulations. This application process is referred to as Gateway 3 and represents another 'hard stop' in the regulatory process. Occupation of a new residential unit in a higher-risk building or part of a higher-risk building before a completion certificate has been granted for that building or part of the building is a criminal offence.

A completion certificate application must include a prescribed list of information regarding the client, the principal dutyholders and a description of the HRB work, including as-built information. This must include a statement from your client confirming that to the best of their knowledge the higher-risk building, as built and including all controlled changes, complies with all applicable requirements of the Building Regulations and that a copy of the golden thread information has been provided to your client's principal accountable person.

The HRB Regulations do not define the meaning of 'as-built information', nor do the regulations prescribe who is responsible for preparing this information. As part of your designer's duties to ensure your client understands their duties you should discuss this aspect of the golden thread and completion certification application with your client. If the new regulatory regime is followed diligently on your project – in particular, the management, recording and approval of all controlled changes following building control approval by the Regulator – the final issue of the approved building design should be representative of the as-built construction. Regular inspections of the construction work during the construction phase will be required to ensure that this is the case and, in some instances, it may also be necessary to procure a survey of the completed construction. You should ensure at the outset of your project that your client allocates sufficient time and resources, including the provision of appropriate professional fees, for such inspections and surveys, either by you, if you are competent to provide this service, or by another suitably qualified professional.

The completion certificate application must also be accompanied by a copy of the building control submission documents approved by the Regulator and a compliance declaration from each principal designer (or sole or lead designer).

Compliance declaration means a document, signed by the principal designer (or sole or lead designer) to which the declaration relates, that includes:
(a) the name, address, telephone number and (if available) an email address of the principal designer (or sole or lead designer);
(b) the dates of their appointment; and
(c) a statement confirming that they fulfilled their duties as a principal designer under Part 2A (dutyholders and competence) of the Building Regulations 2010.

The Regulator must determine a completion certificate application within eight weeks (or within such longer period as the Regulator and your client may agree in writing) beginning with the date the application is received by the Regulator. During this application period, the Regulator must consult with the enforcing authority and sewerage undertaker for the higher-risk building and carry out an inspection of the completed HRB work for the purpose of assessing whether the work complies with all relevant requirements of the Building Regulations.

If your client intends to occupy their building before all of the HRB work is complete on that building, they need to submit a partial completion certificate application to the Regulator. This is similar to a completion certification application and requires generally the same information submittals supplemented with details regarding the part or parts of the building to be occupied prior to overall completion, including a copy of the partial completion strategy. The Regulator must determine a partial completion certificate application within eight weeks, or within such longer period as the Regulator and your client may agree in writing.

Ideally, partial completion should be predetermined by your client prior to commencing the HRB work and a partial completion strategy should be included as part of the building control approval application to the Regulator. If your client proposes to implement a partial completion strategy following building control approval by the Regulator this will be a major change and will require approval by the Regulator prior to the client submitting a partial completion certificate application.

If the Regulator is satisfied that the HRB work is complete and complies with all applicable requirements of the Building Regulations, the Regulator will issue a completion certificate, which is evidence (but not conclusive evidence) of the same. If the Regulator is not satisfied that the HRB work is complete or complies with all applicable requirements of the Building Regulations, the Regulator will issue a rejection notice explaining why the completion certificate application has been rejected.

Receipt of a completion certificate issued by the Regulator under the HRB Regulations will enable your client, or their principal accountable person, to register the building with the Regulator prior to occupation. No part of a completed higher-risk building may be legally occupied unless or until the building has been registered with the Regulator. Registration is dependent on the completion approval process but is separate and additional to it. The regulatory requirements and procedures for this are beyond the scope of this guide, but you can find further details if you need them in Part 4 of the Building Safety Act.

Following completion of your project, as well as providing a copy of the golden thread information, your client is responsible for giving a copy of the *BFLO information* to the *relevant person* no later than the date the construction work is completed.

> **BFLO information** refers to Parts B, F, L and O, and means:
> (a) where Part B of Schedule 1 to the 2010 Regulations imposes a requirement in relation to the work, the fire safety information;
> (b) where paragraph F1(1) of Schedule 1 to the 2010 Regulations imposes a requirement in relation to the work, sufficient information about the building's ventilation system and its maintenance requirements so that the ventilation system can be operated in such a manner as to provide adequate means of ventilation;
> (c) where paragraph L1 of Schedule 1 to the 2010 Regulations imposes a requirement in relation to the work, sufficient information about the building, the fixed building services and their maintenance requirements so that the building can be operated in such a manner as to use no more fuel and power than is reasonable in the circumstances;
> (d) where paragraph L2 of Schedule 1 to the 2010 Regulations applies in relation to the work, sufficient information about

the system for on-site electricity generation in respect of its operation and maintenance requirements so that the system may be operated and maintained in such a manner as to produce the maximum electricity that is reasonable in the circumstances and delivers this electricity to the optimal place for use;

(e) where Part O of Schedule 1 to the 2010 Regulations applies in relation to the work, sufficient information about the provision made in accordance with Part O so that the systems in place further to Part O can be operated in such a manner as to protect against overheating.

Relevant person means:

(a) where, after building work is completed, a building is not a higher-risk building for the purposes of Part 4 of the 2022 Act, the responsible person[31] for the building;

(b) in any other case, the accountable person for the part of the building to which the work relates and the responsible person (if any) for the building.

CHAPTER 8:
BUILDING
SAFETY
DESIGN

Ensuring that any project is designed and built safely and achieves safe outcomes is a collective effort. To that end it is important that the procurement, design, construction and maintenance teams working on your project are working together with a common understanding of what is required of them and of each other. To assist with this endeavour, the British Standards Institute (BSI), sponsored by government, has produced a code of practice, BSI Flex 8670, which provides guidance on the core criteria for building safety applicable to our whole industry.[1]

In this chapter we consider the aspects of the code of practice that are most relevant to designers:

8.1 Fire and life safety
8.2 Structural safety in buildings
8.3 Public health and public safety in buildings
8.4 Construction products and safety critical products

When considering your approach to building safety design it is important to understand the scope and context of your designer responsibilities as well as the extent of your competence. This will ensure that you do not work beyond your knowledge and abilities and that you are able to recognise when it is appropriate or necessary to seek specialist advice to ensure that your design complies with, or exceeds, the minimum relevant requirements of the Building Regulations, standards and supporting statutory guidance.

If your project is, or includes, the design of a higher-risk building, you need to be aware of the impact that this may have on the safety implications set out below and how you should address this in your design.

Public safety issues often pose a greater risk in high-rise buildings and buildings where there are a greater number of residents, particularly vulnerable residents. This is due to the difficulties and increased time it may take to respond to a safety issue and the number of people who may be impacted by the event. For example, in tall buildings the time taken to commence fire-fighting operations may be delayed and escape is likely to be more protracted or difficult, particularly in buildings where people may be asleep, less alert, less mobile or may find it difficult to follow or understand the proposed evacuation strategy.

8.1 Fire and life Safety

To ensure you are competent to design for fire and life safety you need to have a good understanding of:

- the legislative controls for designing for fire safety applicable at the time of designing and/or commencing construction of your project and how these controls contribute to ensuring the fire safety of your project
- the relevant practices relating to managing fire safety and the passive and active technological systems that support these
- the extent of your design role and responsibilities, including interfaces with other designers, and your contribution to the development and application of the fire strategy for your project.

In Chapter 5 we considered the regulatory environment and your statutory duties, including the relevance of Building Regulations and statutory and non-statutory guidance and examples of the regulations and guidance that you need to be familiar with regarding fire safety design.

In Chapter 9 we will consider the principles of fire development and designing for fire safety. This will include the behaviour of fire and fire spread and the measures you need to take to mitigate these in your design and specification, including the application of passive and active fire protection systems and the principles of firefighting and evacuation.

As a designer, you need to establish early on in the design of your project and development of your client's fire strategy the scope of your responsibilities and how these will coordinate and interact with other relevant key individuals, for example, residents, regulators and statutory authorities, fire safety specialists and any statutory dutyholders appointed under the CDM Regulations and Dutyholder Regulations (including the other designers, project consultants and construction contractors and sub-contractors).

8.2 Structural safety in buildings

To ensure you are able to design your project in accordance with the relevant requirements of the Building Regulations, as well as to ensure the structural safety of your project, it is important that you understand the interrelationship of the architectural and structural design of your project. In particular, to understand how your design and specification can impact on and positively contribute to ensuring and maintaining structural safety. This does not require you to have structural expertise, but you do need to understand the basic structural principals relevant to your project in the context of your proposed concept design and how the structural design, fabrication, installation and maintenance interrelates with other aspects of building safety, including fire safety. The extent of structural knowledge you require will depend on the complexity of your project, the sophistication of the structural strategy and the extent to which you have access to and receive competent structural advice.

To ensure your project is structurally safe the structural strategy and design must be undertaken by someone who is appropriately qualified, has suitable experience and is competent to meet the technical demands of your project. For most projects this will require the involvement of a competent structural engineer, or engineers, to oversee the design, manufacture, erection, assembly and inspection of the structure.

Your responsibility as a non-structural designer is to contribute to and assist the structural engineer on your project to establish and maintain the structural safety of that project. You need to assist them to avoid any structural failure that could pose a threat to the safety of people, not only in respect to your project, but also to buildings within the vicinity of your project.

To do this you need to understand:

- the principals of structural design and construction, including how structural systems behave under load and in the event of a fire
- the relevant requirements of the Building Regulations, and relevant codes and standards in relation to the structural stability of the primary and secondary structure and structural fixings

- the requirements for maintaining structural safety during the lifecycle of your project, including requirements for assessment, inspection and maintenance tasks
- how and when to respond to events that can affect structural safety including identifying when it may be necessary to seek competent specialist advice.

These principles are equally important to consider and apply in connection with any temporary conditions during the construction, occupation and maintenance of your project –temporary loads, temporary conditions and temporary works – particularly if your project involves work on an existing occupied building or buildings.

Whilst catastrophic structural failures are rare, their impacts can be severe, including multiple loss of life. Localised structural failure tends to be more common but can still pose a threat to life, for example, the instability of an unrestrained wall or collapse of a parapet wall or balcony. The failure of secondary structural elements can also pose a serious risk, for example, the failure of a guarding or balustrade protecting people from falls from height.

When you are coordinating the design of your project you should ensure that sufficient access is provided to regularly inspect and maintain the primary structural elements of the building.

Serviceability failures can also pose threats to safety, for example, over deflection of a flat roof resulting in ponding of rainwater, imposing loads in excess of the design loading and resulting in the roof collapsing.

Other factors that could result in structural failure include:

- fire
- corrosion
- erosion
- timber decay
- lack of lateral restraint or bracing
- subsidence
- strong winds
- water leaking or flowing below or above ground
- snow or ice accumulation
- driving rains
- chemical reaction
- undermining structural support
- removal of seemingly innocuous loadbearing elements
- weather
- vehicles colliding with the structure.

Ensure sufficient access for maintenance and cleaning of all areas of the building, particularly roof areas to ensure rainwater outlets can be kept clean to avoid excessive build-up of water.

Structural failures tend to be well recorded.[2] It is generally accepted that their causes fall into six categories:

- design or fabrication failures that result in the structure being insufficiently robust or stable to withstand the loads imposed upon it
- construction quality issues, including failing to comply with the structural design or material specification, poor quality workmanship and inadequate supervision or inspection
- use of defective or damaged materials
- substitutions during the procurement or construction of the project that result in inferior systems, products or materials being used
- failure due to fatigue, corrosion, overstress due to movement and exposure to conditions that cause decay
- design and specification inadequacies that fail to take account of all possible unlikely problems such that the as-built structure is insufficiently robust or is exposed to localised structural failure in real-world conditions, for example, as a result of vehicle impact, flooding or explosion.

Ensure that primary structural elements are appropriately designed for their intended use, particularly when introducing new structural elements to existing buildings.

To ensure that the structure of your project is designed with sufficient robustness in recognition of these risks, it is important to understand how your project will perform in the event of a potential failure event or safety occurrence so that steps can be taken in the design to mitigate these risks. For example:

- How long will the structure remain stable in the event of a fire and how long will it remain safe for occupants and firefighters to remain in the building?
- What action should be taken if a structure is potentially compromised by a safety occurrence, for example after a flood or explosion?

You also need to understand the importance of how the planning and design (including design for inspection and maintenance) of your project could contribute to or mitigate potential safety risks to the structure. For example:

- Consider the location and design of storage for fuel and explosive materials that could cause structural failure in the event of an explosion.

Ensure that your project design makes adequate provision for regular inspections and maintenance of all the safety critical construction and installations.

- Consider the height and exposure of external cladding systems to ensure supports and fixings are designed and specified adequately, including provision for future inspection and maintenance.
- Understand how the structure may be vulnerable to deterioration over time and ensure structural elements exposed to weather can be regularly inspected and maintained to maintain their integrity.

Determining the appropriate structural solution for the main structural elements of your project will depend on several factors, for example, the site location, ground conditions, exposure to wind and weather, the potential risk of flooding and the anticipated imposed loads during construction and occupation. These will influence the design of the foundations, the primary structural frame and the load-bearing systems, as well as secondary structural elements and fixings connecting load-bearing elements, including cladding, windows, guarding and balustrades.

As noted above, design for structural safety requires the competence of an engineer, particularly as structural design is becoming increasingly complex and frequently uses advanced analytical techniques to improve structural efficiency. Compliance with the relevant requirements of the Building Regulations and structural codes requires specialist expertise to assess, analyse and address the risk of major structural failure, for example, progressive or disproportionate collapse.[3] Depending on the complexity of your project, the structural design may also require input from a range of specialist contractors, for example, to design connection details, temporary works, façade engineering and fixtures for safety critical elements such as guarding, balustrades and cladding systems. If you are the lead designer on your project it is important that you work closely with the structural engineer and contractor(s) to ensure that all these elements of the design, and the interfaces between them, are properly coordinated and resolved. Again, depending on the complexity of your project, consider discussing with your client whether it may be appropriate to appoint an independent engineer to have general oversight of all the structural design and engineering work, including design, fabrication and installation, to help ensure your project safety outcomes are achieved.

8.3 Public health and public safety in buildings

Whilst fire and structural safety risks are the most likely risks to result in catastrophic failure – where a single event could result in a significant and serious loss of life – public health and public safety risks have the potential to cause serious longer-term harm to the occupants of your project.

Public health risks are those that might give rise to ill health or disease. **Public safety risks** are those that might give rise to injury. As a designer you need to have an awareness of both and understand how your design provides an opportunity to eliminate or mitigate such risks.

The Building Regulations Approved Documents and related British Standards provide useful guidance that will help you to identify and develop a design solution to mitigate public health and public safety risks. Understanding the functional requirements of the Building Regulations that are relevant to your project is essential to ensuring you are able to comply with, or exceed, the minimum requirements of these. This includes understanding how your design interfaces with that of any other designers on your project to ensure that a holistic approach is taken to designing the building safety systems for your project.

Examples of public health and public safety risks that you need to be aware of and mitigate in your design include:

- **Site contaminants**, which may be solid, liquid or gaseous contaminants that could be naturally occurring or result from previous uses of your site, for example polluted ground water, geological factors,[4] radon or methane and chemical or oil spills. Sites that have been subject to industrial or agricultural use can pose a particular risk of contamination. Desktop studies, site surveys and sampling will help to identify the risk of contamination and determine whether specialist advice is necessary to develop a remediation strategy. This may include total or partial removal of contaminants or contaminated material, installing physical barriers such as capping or membranes and/or

Site investigations prior to the construction of your project should include identifying any public health and safety risks, including potential site contaminants.

providing ventilation or sumps to enable the safe dispersal or collection and management of contaminants.

- **Asbestos**, which as we have considered in previous chapters continues to pose a serious health risk in connection with work on existing buildings.
- **Ventilation, damp and moisture**, which if not managed properly will have a detrimental impact on the indoor air quality of your project. Your design must include provision for adequate ventilation, including purge ventilation to help prevent the penetration of moisture into your building, guarding against damp in the habitable spaces as well as interstitial condensation elsewhere in the building. Bear in mind that you may need to balance these requirements with site-specific environmental conditions, for example, pollution or security issues, which will require a coordinated design response to ensure all public health risks are mitigated appropriately.
- **Overheating and heating failure**, which pose serious public health risks if occupants are exposed to excessive or prolonged periods of high

or low temperatures.[5] This is particularly the case for occupants who may be more vulnerable, including the very young and elderly people, or those with underlying health conditions. As well as providing adequate ventilation, your design must include measures to control excessive heat gains, for example from solar gains and/or building services.

- **Water supply, hot water storage, drainage systems and waste**, which need to be designed to prevent them from becoming sources of contamination or pollution within your project, including designing provision for adequate cleaning and maintenance.
- **Gas supply, combustion devices and carbon monoxide**, which require systems to be designed in accordance with relevant regulations, including appropriate provision for ventilation, appropriate supply and discharge points, resistance to decay and damage, leakage detection and access for inspection, servicing and maintenance.
- **Electrical safety and lightning protection**, which pose a serious public safety risk not just from electrocution but also as a potential cause of fire, including as the result of a lightning strike.
- **Guarding, balustrades, staircases and glazing safety**, which as we have considered in previous chapters pose a risk of slips and trips and falls from height, both of which are a frequent source of serious injury in buildings if not designed and specified appropriately.

Appropriate lightning protection is essential to mitigate the serious public safety risks posed by lightning strikes.

The design of systems to address the above safety issues requires careful coordination to ensure that all the specified systems work together and do not compromise one another, or the structural and fire safety measures designed into your building. For example, the design and specification of any service penetrations that breach fire compartmentation should include appropriate fire stopping to maintain the integrity of the compartmentation.

You should also bear in mind that your designer duties include a responsibility to provide information regarding the management, maintenance, installation or replacement of any construction products or building systems you specify. This includes information regarding their specified design and service life as well as the maintenance requirements to comply with product warranties that your client needs to be aware of to ensure the ongoing maintenance and integrity of the building safety systems.

Your design information should include relevant information the building owner/operator will require to maintain the integrity of the safety systems, including ensuring that services installations have adequate fire stopping at junctions with any fire compartmentation.

Depending on the complexity of your project, the level and range of competence required to design these systems will vary. It is likely you will require specialist advice or assistance given the wide range of competence required to address all of the possible public health and public safety issues you will encounter on your project.

8.4 Construction products and safety critical products

As a designer, and particularly if you are appointed as the lead designer and/or principal designer on your project, you have a duty to coordinate your design activities with the rest of your design team and your client's project design team to ensure you, and they, take a holistic approach to building safety.

This means you need to understand the characteristics of the construction materials, products and systems you are specifying and how they interface with one another to perform effectively as a single building system. This includes the importance of procuring, reviewing and understanding the technical literature produced by manufacturers and suppliers, in particular product and system testing, assessment and maintenance information, including identifying and addressing any inadequacies or inaccuracies in that information. Without this you will be unable to assess and mitigate the risks to safety throughout the lifecycle of the building, which may be posed by your design and specification.

When you are considering materials, products and systems as part of your design development you need to establish their appropriateness for your intended use. Think about your project as a holistic system, consider whether the materials, products and systems you are proposing will not only function as you require individually but also whether they will function collectively. Consider the interfaces between details and the interfaces between your design and the design of others to ensure adjacent materials, products and systems work together and are not detrimental to each other's respective performance, integrity and durability.

Your design detailing
and coordination
should include making
sure the materials,
products and systems
you are specifying
are suitable for their
intended purpose
and work together as
a system.

This includes understanding the maintenance requirements of the materials, products and systems you select so that they continue to perform as required. This is not only so you can pass relevant information regarding maintenance back to your client, but also to enable you to ensure that your design makes adequate provision for access to enable proper installation and maintenance of the construction.

To satisfy yourself that your selection and specification of materials, products and systems are correct, you will need to obtain relevant information regarding their performance characteristics, including their durability, in the context of your project. Bear in mind the location and context in which you propose to use each material, product or system. Environmental factors you may need to consider include:

- site exposure, including exposure to any aggressive environmental conditions, for example marine environments
- proximity to the site boundary and any neighbouring buildings, which could pose a risk, for example fire spread from adjoining sites
- size and geometry, for example high wind patterns in and around urban environments and tall buildings.

Obtaining reliable technical product information from manufacturers and suppliers is essential to assisting you with this assessment. Look for independent, third-party accredited and quality assured test information that verifies the product you are considering meets the specified performance criteria you require in the particular circumstances you are proposing. A good example of this is performance certification, supported by a comprehensive test report, provided by a UKAS accredited testing laboratory.[6] Another is CE or UKCA marking, provided that you check the actual performance achieved in relation to the CE or UKCA mark declaration.[7] CE or UKCA marking in and of itself is not confirmation that a minimum performance has been achieved and certified, only confirmation that a product has been tested prior to being placed on the market.

There is also an onus on manufacturers and suppliers to provide comprehensive, accessible and reliable product information.

Under measures introduced by the **Building Safety Act 2022** (BSA), the Office for Product Safety and Standards (OPSS) has responsibility, in its capacity as the National Regulator for Construction Products (NRCP) and working with the Building Safety Regulator and local Trading Standards to provide oversight and national regulation of the construction products market in Great Britain. The NRCP's role includes leading and coordinating market surveillance so that safety concerns relating to construction products are identified and appropriate enforcement action is taken to address these concerns.

Further amendments being introduced under the BSA, that will make changes to the **Construction Product Regulations 2013**, will require that construction products are safe before they are made available on the UK market. The legislation includes a statutory list of 'safety critical' construction products. Safety critical products are those that have been identified as products whose failure could result in serious injury or death. If a construction product is found to be non-compliant the NRCP has the power to recall these products and withdraw them from the market. Manufacturers and suppliers that fail to comply with the Construction Product Regulations will be committing a criminal offence and could face fines and/or imprisonment.

When you are reviewing product data you should always check, and make sure you understand, the field of application for any test certificate and whether the text criteria match those for the material or system you are proposing to use in your design. Laboratory tests can vary considerably from real-life installation, and it is essential that the field of application of the test certificate is relevant to the proposed use. If not, it may be necessary for you to carry out one of several alternative solutions:

- Amend your design to fall within the field of application of the relevant test certificate.
- Choose an alternative product or system that has a field of application suitable for your design.
- Advise your client that it you require an expert to be appointed to advise whether it is possible to justify a proposed variation from the field of application.

- Advise your client to commission design-specific testing with a field of application relevant to your design.

To avoid any problems with the above, wherever possible it is sensible to encourage the manufacturers and suppliers that you work with to commission testing for their products and systems that are relevant to the real-life scenarios you are designing. This will ensure tests with appropriate fields of application are available when you need them to support your design.

CHAPTER 9:
PRINCIPLES OF FIRE SAFETY DESIGN

To ensure your design risk management with respect to fire safety design is effective, you need to understand some of the basic principles of fire science and how construction materials perform in the event of a fire.

In this chapter we consider the aspects of fire and fire safety design with which you need to be familiar:

9.1   Ignition, development and spread of fire
9.2   Fire performance of construction materials
9.3   Design for fire safety

The guidance in this chapter provides a base from which fire safety can be included in the design risk management process. When appropriate, you should seek advice from a suitably qualified and competent fire safety specialist.

9.1 Ignition, development and spread of fire

Ignition of a fire requires an exothermic chemical chain reaction to take place that combines a source of heat, fuel and oxygen. As long as all three of these elements are present, this chemical reaction will sustain a fire. Eliminating one of these elements will extinguish the fire (for example, cooling the seat of the fire, removing the source of fuel or excluding oxygen from the atmosphere of the fire). There are many causes of ignition but typical examples that affect construction sites and/or construction projects include:

- electrical faults
- uncontrolled hot works
- discarded cigarettes that have not been properly extinguished
- arson.

Phases of fire development are used to describe the life of a fire from its ignition to its decay. There are five distinct phases of fire development:

1. The **ignition/incipient phase**, when heat, fuel and oxygen combine in a sustained chemical reaction. This phase typically results in a small fire, and this is when the possibility of extinguishing the fire is greatest.
2. The **growth phase**, when the fire develops as heat from the incipient fire spreads to adjacent combustible materials that act as additional fuel sources and ignite. At this stage, active fire suppression measures – such as sprinklers or mist systems – may extinguish the fire or limit its further growth.
3. The **flashover phase**, when combustible materials within the space containing the fire ignite simultaneously. This normally occurs at temperatures of 500°C or above. Flashover occurs when heat radiating off a buoyant layer of hot smoke, which has spread across the ceiling of a confined room, causes combustible materials in the room to thermally decompose, releasing flammable gases that then ignite.
4. The **fully developed phase**, when the growth phase reaches its peak. The temperature of the fire is at its highest and all combustible

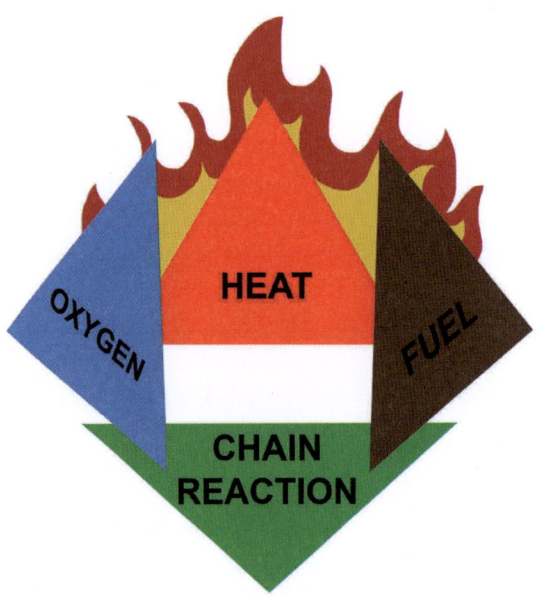

Components of an exothermic chemical chain reaction required to sustain fire.

materials will have been ignited, consuming all available fuel. During this phase, fire safety relies on containment or compartmentation by fire-resisting construction to prevent further fire spread.

5.   The **decay phase**, also known as burnout, when there is a significant decrease in fuel and/or oxygen available to support continued combustion. This results in a fall in the temperature and intensity of the fire, eventually resulting in the fire being extinguished. During this phase, if oxygen is rapidly reintroduced to the oxygen depleted environment (for example, by someone opening a door to the room containing the fire), superheated gasses in the fire will rapidly and explosively burn, creating a backdraft.

Fire spread can occur in three different ways as heat energy from a fire is transferred to adjacent combustible materials, raising their temperature to the point of ignition. These are:

1.   **Convection**, which occurs in fluids and gases when heat transferred from the fire causes the fluid or gas to heat and become less dense and rise, spreading heat away from the vicinity of the fire (for

Phases of fire development and decay.

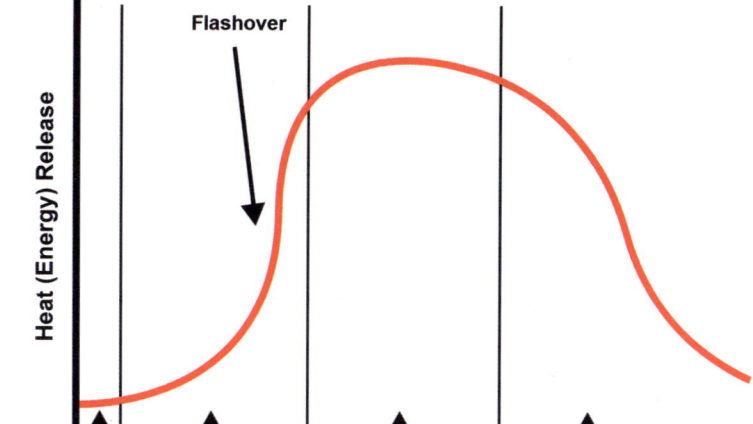

219

example, when air heated by the fire rises and spreads across the ceiling of a room).

2. **Conduction**, which occurs when heat from the fire is transferred between two materials that are in direct contact with each other (for example, when heat is transferred across a solid wall from the side facing on to a fire to the side facing away from the fire).

3. **Radiation**, which occurs when heat from the fire is transferred by electromagnetic waves (for example, the heat you feel being emitted from an electric fire). Radiant heat from a fire can preheat fuels ahead of the fire, increasing the likelihood, speed and intensity of the growth phase.

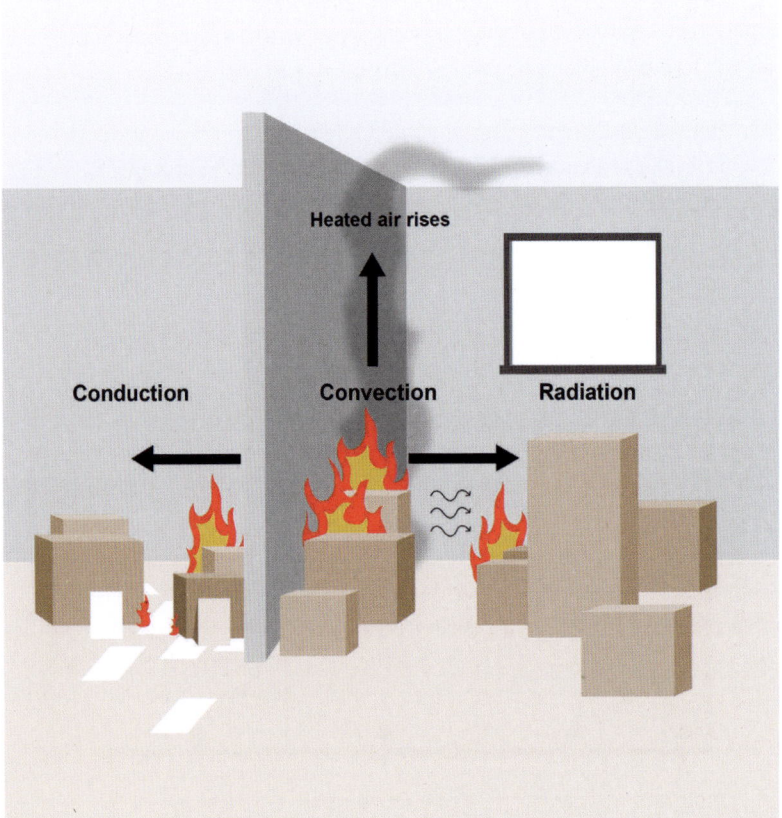

Methods of heat transfer that can lead to fire spread.

9.2 Fire performance of construction materials

The fire performance of construction materials can be classified by two different characteristics: their reaction to fire and their fire resistance.

Reaction to fire is the measure of whether and to what degree a material, product or system will contribute to fire spread. This is done by determining how combustible a material is, how much smoke it will produce once ignited and whether any pieces of the material will sustain combustion.

A material's reaction to fire can be tested, assessed and classified in accordance with BS EN 13501-1. These test results provide a classification from A1 (highest performance i.e. non-combustible) to F (lowest performance i.e. highly combustible). Untested materials cannot be classified.[1]

The classification of a material tested to BS EN 13501-1 will be identified by assessing its gross calorific potential in megajoules per kilogram, temperature rise, mass loss, fire growth rate and duration of sustained flaming.

Combustible materials, products and systems (those rated A2, B, C, D and E) can also be assessed and classified in accordance with BS EN 13501-1 to determine their potential to produce smoke or create flaming droplets. The sub-classifications for smoke production are: s1, s2 or s3, with s1 indicating the lowest production and s3 indicating no limit on smoke production.

Flaming droplets are small particles or pieces of material that, after separating from the source material, continue to burn for a period of time and therefore risk contributing to the spread of fire. The sub-classifications for flaming droplets are: d0, d1 or d2, with d0 indicating no production and d2 indicating no limitation of the production of flaming droplets.

Materials classified as meeting A1 produce very limited smoke and no flaming droplets.

Fire resistance is the ability of a material, component or system to satisfy, for a stated period of time, some or all of the criteria given in a relevant test standard. Fire resistance is measured in minutes, which describes the time elapsed in a standard test rather than real time.

Fire resistance is a measure of one or more of the following:

- Resistance to collapse – a loadbearing element's *loadbearing capacity*. This is denoted **R** in the European classification of the resistance to fire performance.
- Resistance to fire penetration – a measure of a product's *integrity*. This is denoted **E** in the European classification of the resistance to fire performance.
- Resistance to the transfer of excessive heat – a measure of a product's capacity to perform as *insulation*. This is denoted **I** in the European classification of the resistance to fire performance.

Different elements of construction may have combinations of R, E and I. For example, a structural member such as a beam or a column will only have a loadbearing capacity (R), a loadbearing compartment wall or floor will have a loadbearing capacity, integrity and insulation (R, E and I), while a non-loadbearing element such as a partition or a door will only have integrity and insulation (E and I).

The periods of fire resistance given in technical guidance are based on assumptions about the severity of a potential fire and the consequences of any element of a structure failing. Fire severity is estimated in very broad terms based on buildings' uses (their purpose group) and on the assumption that buildings' contents (which constitute the fire load) are similar for buildings in the same purpose group.

Several factors affect the specified period(s) of fire resistance that you will need to specify when designing your project. These include:

- The amount of combustible material per unit of floor area for the type of building you are designing. This is known as the fire load density.
- The height of the top floor of your proposed building above the lowest external ground level. This impacts on evacuation, firefighting and the potential consequences of structural failure.
- The occupancy type, which will determine how quickly and easily your building can be evacuated.
- Whether your project includes any basements. The limited external elevational area of basements limits opportunities for ventilation and access in the event of a fire. This in turn can increase the heat build-up, fire spread and duration of a fire, as well as causing problems for firefighting.
- Whether your building is a single storey and therefore is easy to evacuate and a low risk of structural failure.

Testing and certification are critical in helping you to understand the reaction to fire and fire resistance of any material, product or system you propose to specify, including any limitations on their use.

Appropriate testing information that you should obtain from the manufacturer or supplier prior to specifying a material, product or system should include third-party certification produced following an independent test and assessment by a UKAS accredited testing and certification body.[2] This will provide you with confirmation that the manufacturer's declaration of the fire performance of the materials, product or system have been met under test conditions.

Where it is not possible to obtain third-party certification, an assessment or technical evaluation produced by an accredited fire test laboratory or qualified fire consultant may be used. This will provide you with an expert judgement regarding a material, product or system in lieu of third-party certification.[3] This may be necessary if the size or use of the material, product or system make it impractical to use a standard test method.

You should request a fire test report documenting the results of the British or European Standard fire test that has been carried out by a manufacturer for any fire safety critical material, product or system that you intend to specify. This will provide you with a report that is specific to the fire performance of the specific material, product or system that has been tested, including its finish and colour. You should not rely on indicative or ad hoc tests that have not been carried out in accordance with a British or European Standard fire test; nor should you rely on a test report for a similar product.

Any test evidence you use to determine or demonstrate the fire performance of a material, product or system you intend to specify must be checked to ensure that it is applicable to your intended use. Small differences in detail – such as fixing method, joints, dimensions, the introduction of insulation materials and air gaps (ventilated or not) – can significantly affect the fire performance of a material, product or system.

When you are selecting materials, products or systems for use in the construction of an external wall for a relevant building, you will need to demonstrate that they meet the functional requirements of Building Regulations Part B4.

There are four ways you can do this, which are referred to as routes to compliance:

1. Only specifying materials that have had their fire performance tested and certified with a European Class rating in accordance with **BS EN 13501-1** appropriate for their intended use.
2. Procuring an independent third-party, full-scale fire test to demonstrate that your proposed wall construction meets the performance criteria given in BRE 135 for external walls using full-scale test data from **BS 8414-1** or **BS 8414-2 Fire performance of external cladding systems**.
3. Where it is impractical or not feasible to carry out a full-scale test, procuring an assessment in lieu of a test in accordance with **BS EN 9414 – Fire performance of external cladding systems based on the application of results from BS 8414-1 and BS 8414-2 tests** or **BS EN 15725 – Extended application reports on the fire performance of construction products and building elements**. You should only

rely on this form of assessment when sufficient and relevant test evidence is available to support the assessment, and the assessment has been prepared by an organisation with the necessary expertise (for example, an organisation listed as a notified body in accordance with the European Construction Products Regulations or a laboratory accredited by UKAS for the relevant test standard).[4]

4. Specifying products that are classified as belonging to Classes A 'No contribution to fire' as determined by EU Commission decisions, which remain applicable in the UK and include the majority of common building materials.

In England, the Building Regulations limit the use of combustible materials in the external walls and specified attachments of relevant buildings. Regulation 7 of the Building Regulations defines the scope of relevant buildings, the reaction to fire classification that materials in external walls must meet (European Class A1 or A2, s1, d0) and a list of materials that are exempt from the limitations.

9.3 Design for fire safety

Passive and active fire protection provide the first line of defence to preventing fire growth and spread. The effectiveness of the fire protection measures in a building will also impact on the success of the evacuation strategy and provisions for firefighting.

Passive fire protection refers to static systems that are part of a building's construction (for example, fire-resisting construction used to create fire compartments to contain fire spread). These systems do not require a command signal or intervention to operate.

Examples of passive fire protection include, but are not limited to:

- fire-rated walls and floors
- fire-rated door sets
- fire-rated ductwork and dampers
- cavity barriers, fire stops and fire collars

- alternative means of escape
- protected refuges
- fire-rated firefighting shafts, stairwells and lobbies.

Active fire protection refers to systems that require a command signal or positive action to activate them in the event of a fire. The action may be manual (for example, a portable fire extinguisher) or automatic (for example, a sprinkler system). Active systems are more likely to require regular maintenance and testing and may require building owners/occupiers to be trained in their use and operation. Automatic active fire protection systems also need careful design and specification to ensure they have fail-safe mechanisms and back-up sources of power to confirm they are operational in the event of a fire.

Examples of active fire protection include, but are not limited to:

- smoke, heat and fire detection and alarm systems
- automatic fire suppression systems (for example, sprinkler, mist or CO_2 suppression systems)
- mechanical smoke ventilation systems (for example, automatic opening vents)
- dry and wet risers
- emergency communication systems
- portable fire extinguishers
- emergency lighting.

Passive and active fire protection systems address different aspects of fire protection and are co-dependent, often working together to provide an effective fire protection strategy. For example, fire-rated construction will prevent fire spread whilst an automatic suppression system will limit fire growth. You should consider how this combination – or layers of fire protection – can be used most effectively in the design of your project to produce a comprehensive fire safety strategy that complements your client's requirements for the use, operation and maintenance of the project. As part of your design risk management you should consider the potential impact on your fire strategy of one or more layers of fire protection failing, particularly if you are relying on several active fire protection systems.

Evacuation strategies and the role they play should be a fundamental part of your fire safety strategy and an intrinsic part of the design development of your project.

The evacuation strategy that you adopt will impact on the design choices you make with respect to:

- travel distances
- dead end corridors or single means of escape
- the clear widths of escape routes
- passive and active fire protection measures
- access for firefighting.

When occupants are alerted to fire in a building, they have to make a series of decisions before evacuating. Different people will respond in different ways. The major influence on what they think and do is the nature of the organisational or social unit of which they are a part rather than the design of the building. This broadly relates to the type of building in which occupants experience the emergency. For example, are they familiar with their surroundings at home or work, or are they in a building where they are unfamiliar with the layout? Further guidance regarding the principles of human behaviour during an emergency is provided in Appendix IV.

Understanding human behaviour and responses to a fire, as well as the impact of design on means of warning and evacuation, can inform your design decisions, but you should always balance this with your responsibility to meet your statutory duties and to comply with relevant fire safety design guidance.

The evacuation strategy that is most appropriate for your project will generally depend on its use, size and provision for firefighting. The evacuation strategies that you should consider are:

- **Delayed evacuation**: This includes a remain in place or stay put strategy, which is common in residential developments. This strategy involves limited evacuation of those at direct risk from the fire whilst the majority of occupants remain in the building as the fire is tackled by the fire service. This strategy relies on adequate fire and smoke compartmentation and requires a suitable alternative evacuation plan in case a full evacuation becomes necessary.

- **Phased evacuation**: This strategy is used for buildings where it is not desirable or practical to evacuate all the occupants simultaneously, but delayed evacuation and a stay put strategy is not appropriate (for example, in care homes where the residents require additional time and assistance to prepare for and undertake evacuation). When this approach is used for high-rise residential buildings, the residents most at risk from a fire are evacuated first, including any residents requiring assistance as described in a Personal Emergency Evacuation Plan (see below). Remaining residents are then evacuated in phases, usually two storeys at a time.
- **Progressive evacuation**: This strategy is used when it is not possible for occupants to evacuate the building simultaneously. The strategy involves moving occupants to temporary places of safety within the building where they can remain until they can be evacuated safely. There are two managed strategies:
  1. **Progressive horizontal evacuation**, which involves moving people to an adjoining fire compartment on the same building level. This is a common strategy used in hospitals.
  2. **Zoned evacuation**, which involves moving people away from an affected zone to an adjacent zone. This is a common strategy used in large retail developments.
- **Simultaneous evacuation**: This strategy is adopted when the effects of a fire render it unsafe for occupants to remain in the building. This strategy can be executed in two ways:
  1. **Single-staged evacuation**, which is instigated by a building-wide instantaneous evacuation signal from the fire detection and alarm system.
  2. **Two-staged or double-knock evacuation**, which allows for an investigation period prompted by a localised fire detection and alarm before building-wide evacuation fire alarm systems are activated.

A **Personal Emergency Evacuation Plan (PEEP)** is a bespoke evacuation strategy for an individual who may not be able to reach an ultimate place of safety unaided or within a satisfactory period of time in the event of any emergency. A **Generic Emergency Evacuation Plan (GEEP)** provides a similar form of evacuation strategy for a visitor to a building who may require assistance to evacuate the building.

Provision for firefighting is a functional requirement of the Building Regulations in England, in accordance with Part B5 Access and facilities for the fire service. It includes a requirement to provide reasonable facilities to assist firefighters, as well as reasonable provision to enable firefighters to gain access to your building.

You should consider the aspects of your design that will impact on access and facilities for the fire service from RIBA Plan of Work Stage 2 Concept Design, as part of your fire safety strategy. Guidance regarding appropriate ways to achieve this and how to meet the functional requirements of the Building Regulations is provided in Approved Document B and the British and European standards we considered in Chapter 5.

Understanding the principles of firefighting operations will assist you in developing an appropriate fire safety strategy for your project with your client and their project team. This will be dependent on the nature of your project – the building type and use, its location and size, its occupancy characteristics and your proposed evacuation strategy. The passive and active fire protection measures that you incorporate into your design to ensure the safety of occupants and safe evacuation will also ensure the safety of firefighters. Fire protection measures such as compartmentation, smoke ventilation/extraction and fire suppression systems will provide firefighters with sufficient time to fight a fire.

Additional measures that you should consider incorporating into your building to assist firefighters include providing:

- appropriate vehicular access for firefighting appliances
- fire mains and hydrants
- firefighting shafts (including stairs and lifts)
- basement ventilation.

A detailed commentary regarding these measures is beyond the scope of this guidance. However, if you work on multistorey and/or multi-occupancy buildings or structures you should be aware of how your design may impact on firefighting operations.

Firefighting operations involving multistorey buildings are usually conducted from within the building. The use of external firefighting operations tends to be limited to when fire protection measures have failed, fire spread is out of control and there is a need to contain the fire to prevent it spreading to neighbouring buildings.

This means the vertical circulation core(s) in the building must be designed to fulfil two important and separate functions:

1. Provide a safe smoke-free route out of the building for occupants.
2. Provide a safe route to and from the bridgehead (see below) for firefighters to fight the fire and carry out effective search and rescue operations.

A conflict can arise when these functions must take place at the same time, particularly in single-staircase buildings. Firefighters, often with bulky breathing apparatus, ascending a staircase to reach a fire will hinder the use of the staircase for evacuation. Once the firefighters connect their branch hose(s) to a charged dry riser at the bridgehead, they will need to run the hose(s) from the bridgehead up to the incident floor via the staircase, potentially blocking access for evacuation. Fire doors on the incident floor will then be held open by the charged hose(s) which can result in the staircase filling with smoke, further compromising access for evacuation.

Firefighting in multistorey buildings is typically deployed over several storeys and managed using a 'vertical sectorisation model'. This model is used by firefighters to describe and organise a building into areas of firefighting activity, known as sectors. This sectorisation enables firefighters to manage and control firefighting operations more effectively, particularly when sector commanders cannot be physically present in the

Access to a staircase can be compromised by firefighters and charged hosepipes between the bridgehead and incident floor of a building.

sector, due to smoke or heat for example. These sectors are identified as follows:

Vertical sectorisation model.

- The **bridgehead**, which is normally located at least two storeys below the fire floor, provided that these storeys are clear of smoke, and is the main command point from which firefighting and search and rescue operations are managed on site. To avoid overwhelming the bridgehead with firefighting personnel, a staging area one storey below the bridgehead may be required as a holding area for firefighters awaiting instructions. The bridgehead and staging area refer to locations in the building rather than a sector.
- The **fire sector** is an operational sector and is the main zone of firefighting and rescue operations. The fire sector comprises the storeys from the bridgehead, the fire floor(s) directly involved in the fire – known as the incident floor(s) – and one storey above.
- The **search sector** is an operational sector that is located above the fire sector where search and rescue and ventilation are taking place.
- The **lobby sector** is a support sector and covers the area of operations from the ground floor lobby to the bridgehead.

The vertical sectorisation model illustrated depends on effective compartmentation within a building to contain fire spread. If a compartment fails and fire spreads horizontally or vertically beyond its original compartment, identifying and managing the sectors becomes dynamic and more difficult. This may also result in a change in the firefighting and/or evacuation strategy (for example, moving from a delayed strategy to a simultaneous strategy).

The guidance on fire safety in this chapter has been produced to bridge the gap between technical guidance with which designers will be familiar – including Approved Documents and British Standards – and the principles that underpin fire safety in buildings. Designers should exercise reasonable skill, care and diligence in the performance of their services and should identify when further expertise, in the form of a fire safety specialist, is required.

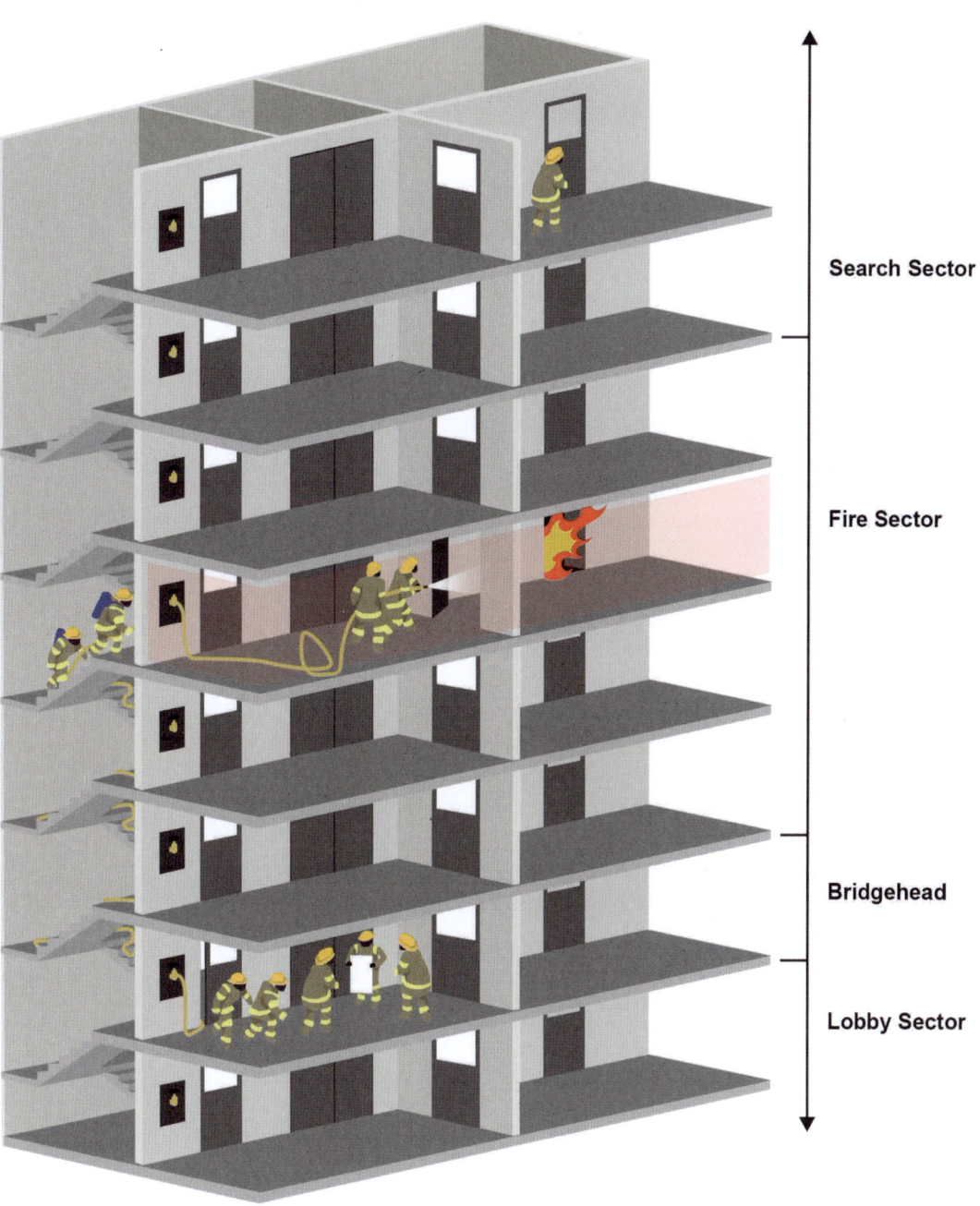

Search Sector

Fire Sector

Bridgehead

Lobby Sector

AFTERWORD

Why does all this matter?

It matters because 72 people lost their lives in the early hours of 14 June 2017, more than 70 people were injured and countless more had their lives changed forever when a domestic fire at Grenfell Tower, west London, burnt out of control, resulting in the most catastrophic residential fire in the UK since the Second World War.

We may not all know the precise details of the circumstances that led to the events on that fateful night, but what we do know is that as a profession we played our part by failing to act on the lessons learnt from a similar fire at Lakanal House, south London, in 2009.

Dame Judith Hackitt's review of the building regulatory system, 'Building a Safer Future', identified a need for systemic change in behaviours and attitudes to safety in the construction industry. Government has responded by regulating for change under the Building Safety Act 2022, introducing new statutory duties for those responsible for the design and construction of buildings. These duties require us to reappraise our approach and attitude to how we deliver our services as designers, principal designers and lead designers.

As a profession, and more importantly as individuals, it is our duty to ensure we embrace this change, to take proper responsibility for the design work we undertake and to remind ourselves of the important role we play in shaping the built environment.

This has to start by ensuring that everyone of us is competent to fulfil the responsibilities entrusted to us to design safe buildings. This includes a responsibility to do everything we can to play our part in improving the construction industry and to collaborate with others to deliver buildings that are safe, both now and in the future.

We owe it to everyone whose life has been impacted by the Grenfell Tower fire to do everything we can to see that such a disaster never happens again.

Dieter Bentley-Gockmann
EPR Architects
June 2023

APPENDIX I: EXAMPLES OF SITE SAFETY SIGNS

No access for
unauthorised persons

Smoking and naked
flames forbidden

No smoking

Do not touch

No access for
pedestrians

Not drinkable

Do not extinguish
with water

No access for
industrial vehicles

Flammable material

Explosive material

Toxic material

Drop

Corrosive material

Biological risk

Overhead load

Obstacles

Industrial vehicles

Danger: electricity

General danger

Low temperature

Laser beam

Oxidant material

Non-ionising radiation

Radioactive material

Eye protection
must be worn

Safety helmet
must be worn

Ear protection
must be worn

Pedestrians must
use this route

Respiratory equipment
must be worn

Safety boots
must be worn

Safety gloves
must be worn

General mandatory sign
(to be accompanied
where necessary by
another sign)

Safety harness
must be worn

Face protection
must be worn

Safety overalls
must be worn

First aid poster

Emergency telephone
for first aid or escape

Stretcher

Eyewash

Emergency exit escape route

Supplementary
directional escape
route

Emergency exit

APPENDIX II: REGULATION 4, HIGHER-RISK BUILDING REGULATIONS

Building control approval applications for HRB work or stage of HRB work

4.—(1) A building control approval application for HRB work must be made in writing, signed by the applicant, and must include—

(a) the name, address, telephone number and (if available) email address of the client;

(b) the name, address, telephone number and (if available) email address of the principal contractor (or sole contractor) and the principal designer (or sole or lead designer);

(c) a statement that the application is made under this regulation;

(d) where HRB work consists of work to an existing building, a description of the existing building including—

   (i) details of its current use, including the current use of each storey;

   (ii) its height as determined in accordance with regulation 5 of the Higher-Risk Buildings (Descriptions and Supplementary Provisions) Regulations 2023(**1**);

   (iii) the number of storeys it has as determined in accordance with regulation 6 of the Higher-Risk Buildings (Descriptions and Supplementary Provisions) Regulations 2023;

(e) a description of the proposed HRB work, including—

   (i) details of the intended use of the higher-risk building, including the intended use of each storey;

   (ii) the height of the higher-risk building as determined in accordance with regulation 5 of the Higher-Risk Buildings (Descriptions and Supplementary Provisions) Regulations 2023;

   (iii) the number of storeys in the higher-risk building as determined in accordance with regulation 6 of the Higher-Risk Buildings (Descriptions and Supplementary Provisions) Regulations 2023;

(iv) the number of flats, the number of residential rooms and the number of commercial units it is proposed the higher-risk building will contain;

(v) the provision to be made for the drainage of the higher-risk building;

(vi) where paragraph H4 of Schedule 1 to the 2010 Regulations imposes a requirement, the precautions to be taken in the building over a drain, sewer or disposal main to comply with the requirements of that paragraph;

(vii) the steps to be taken to comply with any local enactment that applies;

(viii) a statement as to when it is proposed the work is to be regarded as commenced in accordance with regulation 46A (lapse of building control approval: commencement of work) of the 2010 Regulations(**2**).

(2) A building control approval application for HRB work must be accompanied by—

(a) a plan to a scale of not less than 1:1250 showing—

(i) the size and position of the building and its relationship to adjoining boundaries;

(ii) the boundaries of the curtilage of the building, and the size, position and use of every other building or proposed building within the curtilage;

(iii) the width and position of any street on or within the boundaries of the curtilage of the building;

(b)   (i) such other plans as necessary to show that the HRB work

      would comply with all applicable requirements of the building
      regulations(**3**);

  (ii) a competence declaration;

 (iii) a construction control plan;

 (iv) a change control plan;

  (v) a mandatory occurrence reporting plan;

 (vi) a Building Regulations compliance statement;

(vii) a fire and emergency file;

(viii) where the applicant proposes occupation of part of the building
      before completion of the HRB work, a partial completion strategy;

(c) where the application is made by someone on behalf of the client, a
    statement signed by the client confirming they agree to the application
    being made and that the information contained in the application is
    correct.

(3) A building control approval application for a stage of HRB work must
be made in writing, signed by the applicant, and must—

(a) include the information required by paragraph (1);

(b) where the application relates to the first stage of the work—

    (i) be accompanied by a statement ('staged work statement') setting
      out a detailed description of the first stage and of the subsequent
      stages of the project (including an estimate of the time when
      each stage will start);

   (ii) be accompanied by the documents referred to in paragraph (2)
      with the following modifications—

    (aa) the plans referred to in paragraph (2)(b)(i) are such plans as
        necessary to show that the work for the first stage would
        comply with all applicable requirements of the building
        regulations and a summary of plans for the work beyond that
        stage;

    (bb) the Building Regulations compliance statement referred to
        in paragraph (2)(b)(vi) must set out the design principles
        and building standards to be applied to the work for the first
        stage and a summary of the design principles and building
        standards to be applied beyond that stage;

(c) where the application relates to a stage after the first stage—

(i) be accompanied by a statement ('subsequent stages statement') setting out a detailed description of the stage to which the application relates and of the other stages of the project (including an estimate of the time when each remaining stage will start);

(ii) be accompanied by the documents referred to in paragraph (2) with the following modifications—

(aa) the plans referred to in paragraph (2)(b)(i) are such plans as necessary to show that the work comprised in the stage to which the application relates would comply with all applicable requirements of the building regulations and a summary of plans for work beyond that stage;

(bb) the Building Regulations compliance statement referred to in paragraph (2)(b)(vi) must set out the design principles and building standards to be applied to the work comprised in the stage to which the application relates and a summary of the design principles and building standards to be applied beyond that stage.

APPENDIX III: SCHEDULE 1 DOCUMENTS

Higher-Risk Building Regulations: Building control approval application documents

Competence declaration

1.—(1) A competence declaration is a statement, in relation to work, that contains a declaration in relation to—

(a) a principal designer (or sole or lead designer) for HRB work, a stage of HRB work or work to existing HRB,

(b) a principal contractor (or sole contractor) for HRB work, a stage of HRB work or work to existing HRB, and

(c) any other person appointed (A), in relation to the work, as at the date of the building control approval application for HRB work, a building control approval application for a stage of HRB work or the building control approval application for work to existing HRB.

(2) In relation to a principal designer for HRB work, a stage of HRB work or work to existing HRB, the competence declaration must—

(a) state that the client—

(i) has complied with regulation 11E(2)(b) (consideration of past misconduct) of the 2010 Regulations, and

(ii) having taken all reasonable steps, is satisfied that the principal designer fulfils the requirements in regulations 11F(1) and (2) and 11G(1) of the 2010 Regulations, and

(b) include a declaration as to the truth of that statement.

(3) In relation to a principal contractor for HRB work, a stage of HRB work or work to existing HRB, the competence declaration must—

(a) state that the client—

      (i) has complied with regulation 11E(2)(b) (consideration of past misconduct) of the 2010 Regulations, and

      (ii) having taken all reasonable steps, is satisfied that the principal contractor fulfils the requirements in regulations 11F(1) and (2) and 11H(1) of the 2010 Regulations, and

(b) include a declaration as to the truth of that statement.

(4) In relation to a sole contractor, or sole or lead designer, the competence declaration must—

(a) state that the client—

      (i) has complied with regulation 11E(2)(b) (consideration of past misconduct) of the 2010 Regulations, and

      (ii) having taken all reasonable steps, is satisfied that the person fulfils the requirements in regulations 11F(1) and (2) of the 2010 Regulations, and

(b) include a declaration as to the truth of that statement.

(5) In relation to any A, the competence declaration must—

(a) state that the client has been informed by the person appointing A that they have complied with regulation 11E(2)(b) (consideration of past misconduct) of the 2010 Regulations, and

(b) include a declaration as to the truth of that statement.

(6) Any competence declaration must—

(a) be signed by the client, and

(b) where there is more than one contractor for the HRB work, a stage of HRB work or work to existing HRB, include a copy of each record that the client created under regulation 11D(8) or, as the case may be, 11D(9) (principal designer and principal contractor) of the 2010 Regulations.

Construction control plan

2. A construction control plan must set out—

(a) the strategies, policies and procedures the client has adopted for planning, managing and monitoring the HRB work, a stage of HRB work or work to existing HRB so as to ensure compliance with—

    (i) the applicable requirements of the building regulations and to record evidence of that compliance including describing the arrangements the client has adopted to maintain the golden thread information;

    (ii) the duties in Chapter 4 (duties of dutyholders) of Part 2A of the 2010 Regulations;

(b) the strategies, policies and procedures the client has adopted to identify, assess and keep under review the competence of the persons carrying out the HRB work, a stage of HRB work or work to existing HRB or involved in the design of the higher-risk building or design of the building work to the higher-risk building, including the procedures to be followed—

    (i) to determine whether a serious sanction (as defined in regulation 11E of the 2010 Regulations) has occurred in relation to a person to be appointed;

    (ii) to consider any past behaviour in relation to any serious sanction which might call into question the suitability of a person to be appointed;

    (iii) if a person in relation to which a serious sanction has occurred is appointed, to prevent a repeat of the behaviour;

(c) the strategies, policies and procedures the client has adopted to support co-operation between designers, contractors and any other persons involved in the HRB work, a stage of HRB work or work to existing HRB, including the sharing of all necessary information;

(d) a schedule of each appointment which has been made as at the date of the application, giving the name of—

    (i) the person who the client has appointed as the principal contractor (or sole contractor);

    (ii) the person who the client has appointed as the principal designer (or sole or lead designer);

    (iii) any other person (excluding individuals except where they are a sole trader) the client has appointed to work on the project;

(iv) any person (excluding individuals except where they are a sole
   trader) the principal contractor (or sole contractor) has appointed
   to work on the project, and

(v) any person (excluding individuals except where they are a sole
   trader) the principal designer (or sole or lead designer) has
   appointed to work on the project, and a summary of their
   responsibilities;

(e) the policies the client has adopted to review the construction
   control plan.

Change control plan

3.—(1) A change control plan must set out the strategies, policies and
procedures the client has adopted to ensure any controlled change takes
place in accordance with regulation 18 (change control), and to log each
controlled change in accordance with regulation 19 (change control:
record-keeping) including explaining—

(a) how proposed changes will be identified and to whom they must be
   reported;

(b) how the impacts of proposed changes are identified and considered;

(c) in relation to proposed changes, the decision-making procedures
   adopted for agreeing a change including whose advice is to be sought;

(d) how changes are recorded and by when;

(e) the procedure to identify which changes require notification to the
   regulator and which changes require a change control application;

(f) how the effectiveness of the change control plan will be reviewed by
   dutyholders periodically.

(2) In this paragraph 'dutyholders' means the client, the principal
contractor (or sole contractor) and the principal designer (or sole or lead
designer).

Building Regulations compliance statement

4. A Building Regulations compliance statement must set out the
approach taken in designing the proposed higher-risk building or in

designing the work to be carried out on the higher-risk building and the building standards applied, in particular—

(a) the approach taken in relation to each element of the building and the work to ensure compliance with all applicable requirements of the building regulations, and

(b) the reasons for adopting the approach together with an explanation of why the approach is appropriate and ensures compliance with all applicable requirements of the building regulations.

Fire and emergency file

5.—(1) A fire and emergency file must set out—

(a) the matters that were considered when assessing how the building safety risks[1] identified could impact the higher-risk building or the proposed higher-risk building;

(b) the proposals adopted and the approaches taken in relation to designing the proposed higher-risk building or the building work to the higher-risk building to ensure compliance with the applicable requirements of the building regulations relating to the building safety risks;

(c) the measures, strategies and policies it is proposed the owner of the higher-risk building should adopt in order to manage and maintain the higher-risk building or the proposed higher-risk building to ensure anyone in it can be safely evacuated in an emergency, including any assumptions made as to the intended occupiers of the building and their likely characteristics and behaviours.

(2) The measures, strategies and policies referred to in sub-paragraph (1) must include—

(a) a plan which sets out the requirements of the fire and rescue service for the area in relation to access to the higher-risk building and water supply for fire-fighting;

(b) a report which—

  (i) where the HRB work, a stage of HRB work or work to existing HRB has not started, demonstrates how compliance with the

---

(1) See section 120G(5) of the Building Act 1984 for the definition of 'building safety risk'.

applicable requirements of the building regulations relating to the building safety risks is to be achieved;

(ii) where the HRB work, a stage of HRB work or work to existing HRB is completed, demonstrates how compliance with the applicable requirements of the building regulations relating to the building safety risks was achieved.

Partial completion strategy

6. A partial completion strategy must explain—

(a) the proposals adopted in designing for occupation of each part of the building or the proposed building to be completed to ensure compliance with the applicable requirements of the building regulations;

(b) the measures, strategies and policies it is proposed the owner of the building should adopt in order to manage and maintain each such part of the building or the proposed building;

(c) any assumptions made in those measures, strategies and proposals as to the intended occupiers of each such part of the building or the proposed building and their likely characteristics and behaviours, and the intended management or maintenance of each such part of the building or the proposed building.

APPENDIX IV: PRINCIPLES OF HUMAN BEHAVIOUR DURING AN EMERGENCY

Warning and escape in case of fire

Our interventions are not only shaped by their context, purpose and construction methodology, but also by overarching legislative requirements. In the event of a fire, the regulations specifically set out several requirements including the need to provide the appropriate provision for communicating or warning of fire and the appropriate means of escape in case of fire.

There are several approaches that can be followed in order to demonstrate compliance with these regulatory requirements, for example, using approved document guidance or the application of fire safety engineering principles. These approaches provide guidance based on the time taken for an individual to move from their initial position to a place of safety in an emergency. The time taken is generally expressed as a travel distance to enable a designer to set out an escape route. Calculations and modelling using computational tools and qualitative design review can also be used to benchmark designs against regulations and other requirements and recommendations.

These routes to compliance do not account for the actual actions of individuals before and during a fire, which can delay their exit. There are however several models that begin to address behaviours performed by occupants during fires, such as EXODUS, which focuses on five core interacting sub-models – Occupant, Movement, Behaviour, Toxicity and Hazards – and CRISP (Comparison of Risk Indices by Simulation Procedures), a risk assessment model which focuses on the physical and chemical process of fire in relation to the behaviours of occupants attempting to escape or suppress the fire.

Human behaviour in an emergency

In order to recognise the impact of human behaviours in fire, and the effect of active and passive layers of fire protection, we need to understand and appreciate how these two fundamental aspects directly impact occupant safety. Human behaviour specifically focuses upon patterns and actions developed through an individual's decision-making process, which is dependent on the type of building and design related to aid an occupant's safety should they decide to, or need to, stay put or leave.

Understanding human behavioural patterns and response in a fire or other emergencies, as well as the impact of design on means of warning and escape, can begin to inform design decisions, but should not unfavourably impact upon your responsibilities to meet baseline legislative requirements (see Chapter 5).

When alerted to a fire in a building the users have to make a series of decisions. Although different people may respond in different ways, the major influence on what they think and do, is not the design details, but the nature of the organisational or social unit they are part of. This broadly relates to the type of building in which they experience the emergency.

Decision making in an emergency

This model identifies the process of decision-making in an emergency, such as a fire, through several stages and highlights various routes that illustrate behaviour patterns between three distinct stages:

249

- **Interpret**: The process is triggered when an individual is presented with initial cues ('receive information'), upon which there is a decision to investigate further ('investigate') or misinterprets these initial cues ('ignore').
- **Prepare**: Once a fire has been identified, the individual will attempt to seek further information ('explore'), inform others ('instruct') or remove themselves from the situation ('withdraw').
- **Act**: The final stage, the individual will deal with the fire ('fight'), interact with others ('warn') and either leave ('evacuate') or stay ('wait').

This model can be used to describe the decision process during a fire, where the organisational structure of different building types creates a set of social rules that inform a pattern of occupant's decisions when seeking information and taking action during a fire. Using different building types, the following examples show how the decisions made when interpreting, preparing and acting, during a fire, relates to organisational and social structures.

The final stage of acting can be hindered through the design of buildings. A typical plan, which is repetitious across the number of stories of

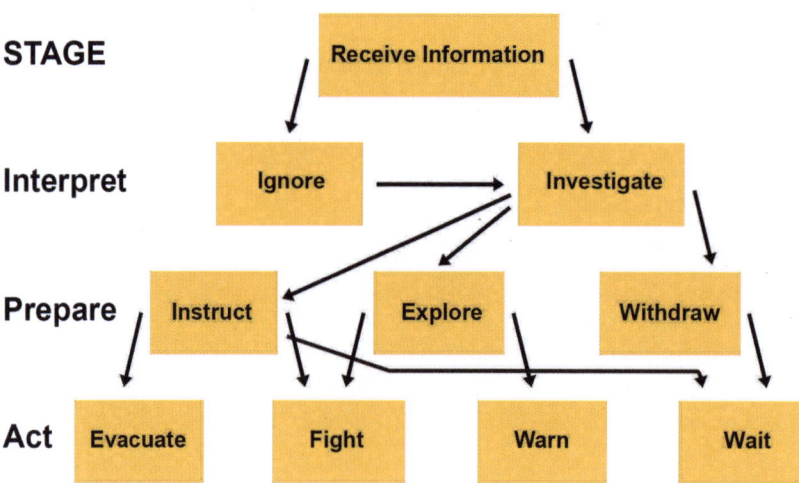

Model of human behaviour in an emergency.[1]

a building, or simple plan where it is easy to orientate yourself, such as sightlines to escape routes (stairs and lifts), avoids confusion to occupants trying to leave. In complex layouts where the plan form changes between storeys, this can cause poor awareness of access and escape routes requiring careful wayfinding in the design. Typically, if occupants can find their way around building under normal conditions, then they will be able to in an emergency.

Human behaviours in different building types and organisational structures

Homes

Occupants generally misinterpret or ignore cues of a fire and will only investigate further where cues persist. Each occupant tends to investigate the situation themselves to confirm the presence of a fire, even if confirmed by other occupants. Before evacuating there may be some attempt to fight the fire, which is particularly characteristic of domestic fires as opposed to other building types.

Multi-occupancy residential

Occupants often receive ambiguous cues, which are likely to be followed by a process of misinterpretations. Occupants may find it confusing and difficult to understand if they are the prime discoverer of a fire, or one of many individuals with a similar experience. During evacuation, most individuals return to their flat before evacuating the building and their awareness of the likelihood of meeting others is demonstrated by their actions of dressing and gathering valuables, delaying their exit.

Hotels

Similar to multi-occupancy residential buildings, guests receive, misinterpret and investigate the cues of a fire. Invariably, guests would inform a member of staff and seek information from an authoritative source in preparation to act. This influences the actions of the guests, where instructions are relayed from hotel staff, primarily from guests telephoning the hotel reception. Dressing, gathering valuables and packing are common in these emergency situations, and guests would usually evacuate by the available stairs, even if hindered by smoke.

Offices

In offices, occupants are usually familiar with their surroundings (including various escape routes from frequent fire safety drills) and have a clearer knowledge of their organisational structure and hierarchy. The process of decision-making is directed around the organisational structure, and although occupants will investigate cues and may confirm the effects of a fire, or potential fire, they will alert senior members of staff rather than explore further.

Public buildings

In public buildings, there is a reliance on people in authority (usually staff) for information. Many people in the early stages of fire growth will tend to continue to carry out ineffective actions, delaying their own response to the early warnings. Staff may investigate and try to address the fire themselves, even when they acknowledge that helping people to evacuate the building is the most effective option. Usually, when appropriate information is provided to the public, evacuation is frequently rapid.

Hospitals

Detection and investigation of a fire takes place relatively early when compared with other occupancies. This is due to the more general spread of people throughout the building and the fact that there is always somebody awake and on duty. Routes of escape are carefully pre-planned and during a fire they are relayed by senior staff and assisted by junior staff, removing decision-making from patients and visitors.

Key patterns of human behaviour

Research findings since the early 1990s have highlighted a series of familiar behavioural patterns and actions of occupants during a fire. In most cases, occupants continue to act in a way that they know is appropriate, for example, they leave the way they came in. These propositions not only take account of the objective physical environment, but also of people's knowledge of a building layout and the information available about a fire threat.

- In the early stages of an emergency, occupants seek to obtain clear information and/or direction from an authoritative figure to tell them what the emergency is and what they are to do.
- Deaths in large-scale fires attributed to 'panic' are far more likely to have been caused by delays in people receiving information about a fire.
- Fire alarm sirens cannot always be relied upon to prompt people to immediately move to safety.
- The start-up time (i.e. people's reaction to an alarm) is just as important as the time it takes to physically reach an exit.
- Much of the movement in the early stages of fires is characterised by investigation, not escape.
- As long as an exit is not seriously obstructed, people have a tendency to move in a familiar direction (the way they came in) even if further away, rather than to use a conventional unfamiliar fire escape route.
- Individuals often move towards and with group members and maintain proximity as much as possible with individuals to whom they have emotional ties.
- Fire exit signs are not always noticed (or recalled) and may not overcome difficulties in orientation and wayfinding imposed on escapees by the architectural layout and design of an escape route.
- People are often prepared, if necessary, to try to move through smoke.
- People's ability to move towards exits may vary considerably (e.g. a young fit adult as opposed to a person who is elderly or who has a disability).[2]

NOTES

Chapter 1

1   There are two types of survey for asbestos: a management survey and a refurbishment/demolition survey. The purpose of a management survey is to manage asbestos during the normal occupation and use of a site and must identify any asbestos that could be damaged or disturbed by normal activities, including foreseeable maintenance. A refurbishment/demolition survey is required where the site – or part of it – needs refurbishment or demolition to ensure that such work is done by the right contractor in the right way and must locate and identify all asbestos before any work begins.

2   The Disclosure and Barring Service is a government agency that helps employers make safer recruitment decisions by processing and issuing DBS checks for England, Wales, the Channel Islands and the Isle of Man. DBS also maintains the adults' and children's Barred Lists and makes considered decisions as to whether an individual should be included on one or both of these lists and barred from engaging in regulated activity.

3   Rigger boots, whilst widely available and regularly used by operatives on site, do not provide ankle support and are not safety rated for use on construction sites.

Chapter 3

1   Health and Safety Executive, 'Why is asbestos dangerous?', *Health and Safety Executive*, www.hse.gov.uk/asbestos/dangerous.htm [accessed 17 January 2020].

2   Health and Safety Executive, 'Construction statistics in Great Britain, 2022', *Health and Safety Executive – Annual Statistics*, 2022, www.hse.gov.uk/statistics/industry/construction.pdf [accessed 26 January 2023].

3   Ibid.

4   Drones present a hazard to aviation and their use is regulated in law by the Air Navigation Order. More information is available on the government's website www.gov.uk/government/news/drones-are-you-flying-yours-safely-and-legally [accessed 26 January 2020].

5   The Confined Spaces Regulations 1997 impose legal duties on employers and self-employed people to ensure that work that needs to be carried out in confined spaces is managed safely and in accordance with the regulations.

6   Health and Safety Executive, 'Asthma', *Health and Safety Executive*, www.hse.gov.uk/asthma [accessed 25 February 2020].

7   Health and Safety Executive, 'Silicosis', *Health and Safety Executive*, www.hse.gov.uk/lung-disease/silicosis.htm [accessed 25 February 2020].

8   Health and Safety Executive, 'Chronic obstructive pulmonary disease (COPD)', *Health and Safety Executive*, www.hse.gov.uk/copd/index.htm [accessed 25 February 2020].

9   The Control of Substances Hazardous to Health Regulations 2002 are intended to prevent or control exposure to substances that are hazardous to health and impose duties on employers to identify and control the use of such substances.

10 Health and Safety Executive, 'Construction statistics in Great Britain, 2022', *Health and Safety Executive – Annual Statistics*, 2022, www.hse.gov.uk/statistics/industry/construction.pdf [accessed 26 January 2023].

11 NHS, 'Leptospirosis (Weil's disease)', *NHS*, 2020, www.nhs.uk/conditions/leptospirosis [accessed 25 February 2020].

12 Health and Safety Executive, 'Harmful Micro-Organisms: Leptospirosis/Weil's Disease from Rats', *Health and Safety Executive*, www.hse.gov.uk/construction/healthrisks/hazardous-substances/harmful-micro-organisms/leptospirosis-weils-disease.htm [accessed 25 February 2020].

13 RHS, 'Giant Hogweed', *RHS*, www.rhs.org.uk/advice/profile?pid=458 [accessed 17 January 2020].

14 GOV.UK, 'Psittacosis', *GOV.UK*, 2017, www.gov.uk/guidance/psittacosis [accessed 17 January 2020].

15 Health and Safety Executive, 'Extrinsic Allergic Alveolitis', *Health and Safety Executive*, www.hse.gov.uk/lung-disease/extrinsic-allergic-alveolitis.htm [accessed 17 January 2020].

16 NHS, 'Lyme disease', *NHS*, 2018, www.nhs.uk/conditions/lyme-disease [accessed 17 January 2020].

17 NHS, 'Insect Bites and Stings – Symptons', *NHS*, 2019, www.nhs.uk/conditions/insect-bites-and-stings/symptoms[accessed 17 January 2020].

18 NHS, 'Snake bites', *NHS*, 2019, www.nhs.uk/conditions/snake-bites [accessed 17 January 2020].

Chapter 4

1 'Reasonably practicable' means balancing the level of risk against the measures needed to control the risk in terms of money, time or trouble. You do not need to take action if it would be grossly disproportionate to the level of risk.

2 In Chapter 5 we consider the legislation and industry guidance that describes what competence means for architects and designers, and how these impact your role within the design team.

3 Cognitive diversity is defined as the differences in our thought and problem-solving processes. Our differences in terms of culture, background, experiences and personalities are core to diverse thought. Bringing together people who think differently from one another can create conversations that stimulate new ideas and innovation.

4 S.I. No. 2023/907 The Higher-Risk Buildings (Management of Safety Risks etc) (England) Regulations 2023 which came into force at the same time as Section 83 of the Building Safety Act 2022.

5 Health and Safety Executive, 'Fire safety in construction', *Health and Safety Executive*, www.hse.gov.uk/pubns/books/hsg168.htm [accessed 24 April 2022].

6 Detailed guidance on the principles of fire safety is beyond the scope of this publication.

7 Health and Safety Executive, 'Working at height', *Health and Safety Executive*, www.hse.gov.uk/toolbox/height.htm [accessed 24 April 2022].

8 Health and Safety Executive, 'Musculoskeletal disorders', *Health and Safety Executive*, www.hse.gov.uk/msd/index.htm [accessed 24 April 2022].

9 Health and Safety Executive, 'Construction dust', *Health and Safety Executive*, www.hse.gov.uk/pubns/cis36.pdf [accessed 24 April 2022].

10   Health and Safety Executive, 'Assessing the slip resistance of flooring', *Health and Safety Executive*, www.hse.gov.uk/pubns/geis2.pdf [accessed 24 April 2022].

11   Test data should be provided by an UKAS certified testing laboratory, or similar, and undertaken in accordance with HSE and UKSRG recommendations using a pendulum test in accordance with BS EN 16165.

12   Construction Industry Training Board, 'CDM2015 – Industry guidance for Designers', *Construction Industry Training Board 2015*, www.citb.co.uk/media/ndlbnb5v/cdm-2015-designers-interactive.pdf, 2015 [accessed 24 April 2022].

13   HM Government, 'The Management of Health and Safety at Work Regulations 1999: Schedule', legislation.gov.uk, www.legislation.gov.uk/uksi/1999/3242/schedule/1/made [accessed 24 April 2022].

14   A safety management plan should include documented details of arrangements to implement, control, monitor and review measures required to manage the hazard identified that may pose a risk to employees.

Chapter 5

1   The United Kingdom (UK) comprises England, Northern Ireland, Scotland and Wales. Although the UK is a unitary sovereign state, Northern Ireland, Scotland and Wales have an increasing degree of legislative autonomy through the process of devolution. Legislation enacted across Great Britain applies to England, Scotland and Wales.

2   The HSWA is applicable in England and Wales, but all functions of the Crown under the Act are devolved to the Welsh Assembly for projects undertaken in Wales.

3   The Building Act is applicable in England and Wales but all functions of the Crown under the Act are devolved to the Welsh Assembly for projects undertaken in Wales.

4   The extent to which the BSA applies to the UK varies. Some parts of the BSA apply to all of the UK, the application of some parts is limited to England and Wales only and further parts are applicable to England only.

5   Responsibility for Building Regulations in Wales is devolved to the Welsh Assembly.

6   Details of the classes of buildings exempt from parts of the Building Regulations are set out in Schedule 2 of the Building Regulations, www.legislation.gov.uk/uksi/2010/2214/schedule/2/made [accessed 1 May 2022].

7   As defined in Regulation 3 of The Regulatory Reform (Fire Safety) Order 2005.

8   HM Government, 'Building Safety Act 2022', legislation.gov.uk, www.legislation.gov.uk/en/ukpga/2022/30/contents [accessed 19 May 2022].

9   Hackitt, Dame Judith, 'Independent Review of Building Regulations and Fire Safety: Final Report', GOV.UK, 2018, www.gov.uk/government/publications/independent-review-of-building-regulations-and-fire-safety-final-report [accessed 1 May 2022].

10   S.I. 2023 No. 275 The Higher-Risk Buildings (Descriptions and Supplementary Provisions) Regulations 2023 which came into force on 06 April 2023.

11   S.I. 2023 No 906 The Building (Approved Inspectors etc. and Review of Decisions) (England) Regulations 2023 which came into force on 01 October 2023.

12   S.I. 2023 No. 911 The Building Regulations etc. (Amendment) (England) Regulations 2023 which came into force on 01 October 2023.

13   S.I. 2023 No. 909 The Building (Higher-Risk Buildings Procedures) (England) Regulations 2023 which came into force on 01 October 2023.

14   Explanatory Notes were introduced in 1999 and Explanatory Memoranda were introduced in 2004. Both can be accessed in respect of legislation laid before parliament after these dates via legislation.gov.uk.

15   England and Wales each have their own Approved Documents, tailored to suit the detail of the Building Regulations as they apply to each jurisdiction.

16   Ministry of Housing, Communities & Local Government, 'Building (Amendment) Regulations 2018: frequently asked questions', GOV.UK, 2020, www.gov.uk/government/publications/

building-amendment-regulations-2018-frequently-asked-questions/building-amendment-regulations-2018-frequently-asked-questions [accessed 1 May 2022].

17   Health and Safety Executive, 'Building Control: An Overview of the new regime Gateways 2 and 3 – application to completion certificate', *Health and Safety Executive*, www.hse.gov.uk/building-safety/building-control/regime-overview.htm [accessed 31 August 2023].

18   European Standards (ENs) are British Standards (BSs) that have been ratified by one of three European Standardisation Organisations. ENs continue to be applicable to projects undertaken in the UK following the UK's departure from the European Union.

19   Including products incorporated within building elements with the exception of power, control and communication cables which are covered by EN 13501-6.

20   The fire performance of construction products may also be classified in the UK in accordance with BS 476 fire tests on building materials and structures, which is referred to as the national classification. Table B1 of Approved Document B transposes the fire classifications from BS EN 13501-1 to the national classes set out in BS 476. The national classifications do not automatically equate with the transposed classifications in BS EN 13501-1 and therefore products cannot typically assume a European class unless they have been tested accordingly.

21   Health and Safety Executive, 'Legal Status of HSE guidance and ACOPs', *Health and Safety Executive*, www.hse.gov.uk/legislation/legal-status.htm [accessed 1 May 2022].

22   Centre for Window and Cladding Technology and Society of Façade Engineering, 'Technical guidance for interpretation in relation to the external walls and specified attachments of Relevant Buildings in England', Issue 2, July 2023, www.cwct.co.uk/pages/cwct-sfe-fire-guidance [accessed 31 August 2023].

23   HM Government, 'Architects Act 1997', legislation.gov.uk, www.legislation.gov.uk/ukpga/1997/22/contents [accessed 2 May 2022].

24   Architects Registration Board, 'Architects Code: Standards of Conduct and Practice', Architects Registration Board, www.arb.org.uk/architect-information/architects-code-standards-of-conduct-and-practice/ [accessed 2 May 2022].

25   Architects Registration Board, 'Architects Code: Advisory Notes, Standard 2: Competence', Architects Registration Board, www.arb.org.uk/architect-information/architects-code-standards-of-conduct-and-practice/advisory-notes/competence/ [accessed 2 May 2022].

26   Royal Institute of British Architects, 'RIBA Code of Professional Conduct', Architecture.com, www.architecture.com/knowledge-and-resources/resources-landing-page/code-of-professional-conduct [accessed 2 May 2022].

27   'Reasonable steps' may require you to do more than is strictly required by law and regulations.

28   Royal Institute of British Architects, 'The Way Ahead', Architecture.com, 2021, www.architecture.com/knowledge-and-resources/resources-landing-page/the-way-ahead [accessed 2 May 2022].

Chapter 6

1   A domestic client is defined in the CDM Regulations as a client for whom a project is being carried out which is not in the course or furtherance of a business of that client.

2   The Construction Industry Training Board (CITB) has published guidance regarding all duties holders' responsibilities under the CDM Regulations that can be accessed on their website at CITB, 'Construction (Design and Management) Regulations', *CITB*, www.citb.co.uk/about-citb/partnerships-and-initiatives/construction-design-and-management-cdm-regulations/cdm-regulations/ [accessed 22 January 2020].

3   Anyone that prepares, modifies, arranges for or instructs design work is deemed to be a designer under the CDM Regulations, including clients and contractors, and assumes the duties of a designer under the CDM Regulations. This includes anyone that may not be part of the project team but still instructs you in connection with your design. For example, if a

planning officer instructs you to modify your design and that instruction does not directly relate to ensuring compliance with planning policy, that planning officer will be deemed to be a designer under the CDM Regulations.

4 For complex or unusual projects, it may be beneficial for the client to appoint a health and safety specialist to support the design team. Some clients also choose to appoint a health and safety or CDM adviser to advise them regarding their client duties and may also choose to appoint the same specialist as the principal designer, but the duties of the two roles are distinct and should not be confused.

Chapter 7

1 Hackitt, Dame Judith, 'Independent Review of Building Regulations and Fire Safety: Hackitt review', GOV.UK, 2018, www.gov.uk/government/collections/independent-review-of-building-regulations-and-fire-safety-hackitt-review [accessed 19 May 2022].

2 HM Government, 'Building Safety Act 2022', legislation.gov.uk, www.legislation.gov.uk/en/ukpga/2022/30/contents [accessed 19 May 2022].

3 Ministry of Housing, Communities & Local Government, 'Building a safer future: proposals for reform of the building safety regulatory system', GOV.UK, 2019, www.gov.uk/government/consultations/building-a-safer-future-proposals-for-reform-of-the-building-safety-regulatory-system [accessed 19 May 2022].

4 HM Government, 'Architects Act 1997', legislation.gov.uk, www.legislation.gov.uk/ukpga/1997/22/contents [accessed 19 May 2022].

5 HM Government, 'Building Act 1984', legislation.gov.uk, www.legislation.gov.uk/ukpga/1984/55/contents [accessed 19 May 2022].

6 HM Government, 'Defective Premises Act 1972', legislation.gov.uk, www.legislation.gov.uk/ukpga/1972/35 [accessed 19 May 2022].

7 HM Government, 'The Regulatory Reform (Fire Safety) Order 2005', legislation.gov.uk, www.legislation.gov.uk/uksi/2005/1541/contents/made [accessed 19 May 2022].

8 'Building function' means (a) any function of the regulator under, or under an instrument made under, this Act or the Building Act 1984; (b) any prescribed function of the regulator; (c) any function of the regulator under the Health and Safety at Work etc. Act 1974 so far as relating to a function within (a) or (b).

9 'Relevant persons' means residents of higher-risk buildings, owners of residential units in such buildings, persons who are accountable persons and persons upon whom duties are imposed by virtue of paragraph 5B of Schedule 1 to the Building Act 1984 (dutyholders).

10 'Built environment industry' means persons carrying on, for business purposes, activities connected with the design, construction, management or maintenance of buildings, and employees of such persons; and references to a person 'in' the industry are to any such person or employee.

11 'Registered building inspector' has the meaning given by section 58A of the Building Act 1984.

12 A 'voluntary occurrence reporting system' is a system to facilitate the voluntary giving of information about building safety to the person who operates the system.

13 The Building Regulations Advisory Committee for England was established under section 14 of the Building Act 1984 to advise the Secretary of State in connection with the Building Regulations. This committee is abolished by Section 9(3) of the Building Safety Act 2022.

14 'Authorised officer' means a person in respect of whom an authorisation under Section 22 of the Building Safety Act 2022 is in force.

15 'Local authority' means a district council or relevant unity authority, a London borough council, the Common Council of the City of London, the Sub-Treasurer of the Inner Temple, the Under Treasurer of the Middle Temple, or the Council of the Isles of Scilly.

16 If commencement for your project will be defined by completion of 15% of the works, you need to agree with your client the works that will constitute this completion and confirm the

details as part of the building control approval application so that they can be agreed with the building control authority prior to works starting on site.

17 Section 2, The Higher-Risk Buildings (Descriptions and Supplementary Provisions) Regulations 2023.
18 Section 7, The Higher-Risk Buildings (Descriptions and Supplementary Provisions) Regulations 2023.
19 S.I. 2023 No. 275 The Higher-Risk Buildings (Descriptions and Supplementary Provisions) Regulations 2023.
20 Further information regarding how to measure the height of a higher-risk building including explanatory diagrams is provided as part of the government's online guidance https://www.gov.uk/guidance/criteria-for-being-a-higher-risk-building-during-the-occupation-phase-of-the-new-higher-risk-regime [accessed 25 September 2023].
21 The BSA confers power on Welsh Ministers to define 'higher-risk buildings' in Wales.
22 Planning Gateway 1 was introduced on 1 August 2021 under the Town and Country Planning (Development Management Procedure and Section 62A Applications (England) (Amendment) Order 2021. The gateway process is only relevant to higher-risk buildings.
23 As defined in section 120G(5) of the Building Act 1984.
24 The 'enforcing authority' within the meaning of article 25 of the Regulatory Reform (Fire Safety) Order 2005.
25 Section 30A of the Building Act 1984 as amended by Section 37 of the Building Safety Act 2022.
26 Regulation 9, The Building (Higher-Risk Buildings Procedures) (England) Regulations 2023.
27 The definition of Major Change differs in relation to HRB work (or a stage of HRB work) and work to existing HRBs. The definition provided is for HRB work. The definition for work to existing HRBs is provided in Regulation 26 (1)(b), The Building (Higher-Risk Buildings Procedures) (England) Regulations 2023.
28 Regulation 25, The Building (Higher-Risk Buildings Procedures) (England) Regulations 2023.
29 Regulation 19, The Building (Higher-Risk Buildings Procedures) (England) Regulations 2023.
30 Regulation 31, The Building (Higher-Risk Buildings Procedures) (England) Regulations 2023.
31 'Responsible person' has the meaning given in Article 3 of the Regulator Reform (Fire Safety) Order 2005.

Chapter 8

1 BSI Flex 8670: v3.0 2021-04, BSI Standards Limited 2021.
2 Collaborative Reporting for Safety Structures UK (CROSS-UK) provides a confidential reporting system which allows professionals working in the built environment to report on fire and structural safety issues, www.cross-safety.ork/uk.
3 Progressive or disproportionate collapse is where the failure of one part of a building's structure leads to other parts of the structure collapsing.
4 Geological factors may include naturally occurring heavy metals such as cadmium and arsenic and naturally occurring gases such as methane (which poses a risk of explosion and fire) or radon (which could cause cancer).
5 There is a serious risk to health from people spending extended periods of time living with internal temperatures below 13°C.
6 UKAS is the National Accreditation Body for the United Kingdom, www.ukas.com.
7 UKCA (UK Conformity Assessed) marking is used for certain products placed on the market in Great Britain (England, Wales and Scotland) and is being introduced to replace CE Marking which will no longer be recognised for construction products after 30 June 2025. An additional mark, UKNI, may be used in Northern Ireland.

Chapter 9

1   Materials covered by the Classification Without Further Testing (CWFT) process can be found by accessing the European Commission's website eur-lex.europa.eu/.
2   United Kingdom Accredited Service.
3   Details of best practice in conducting assessments and technical evaluations is provided in the Passive Fire Protection Forum's 'Guide to undertaking technical assessments of the fire performance of construction products based on fire test evidence, 2021' https://www.firesectorfederation.co.uk/wp-content/uploads/2021/02/Guide-to-Undertaking-Technical-Assessments-of-the-Fire-Performance-of-Construction-Products-Based-on-Fire-Test-Evidence-2021-1-2.pdf [accessed 6 June 2023].
4   2003/424/EC 'Commission Decision of 6 June 2003 amending Decision 96/603/EC establishing the list of products belonging to Classes A 'No contribution to fire' provided for in Decision 94/611/EC implementing Article 20 of Council Directive 89/106/EEC on construction products', notified under document number C(2003) 1673, https://www.legislation.gov.uk/eudn/2003/424/pdfs/eudn_20030424_adopted_en.pdf [accessed 6 June 2023].

Appendix IV

1   Canter, David, 'An Overview of Behaviour in Fires', in *Psychology in Action* (Dartmouth Publishing Company, London, 1996), pp.159–88, http://eprints.hud.ac.uk/9228/1/CANTER_159.pdf [accessed 7 May 2023].
2   Sime, Johnathon, 'Human Behaviour in Fires', Building Use and Safety Research Unit, School of Architecture, Portsmouth Polytechnic, 1991.

INDEX